David A. Elizondo, Agusti Solanas, and Antoni Martínez-Ballesté (Eds.)

Computational Intelligence for Privacy and Security

Studies in Computational Intelligence, Volume 394

Editor-in-Chief
Prof. Janusz Kacprzyk
Systems Research Institute
Polish Academy of Sciences
ul. Newelska 6
01-447 Warsaw
Poland
E-mail: kacprzyk@ibspan.waw.pl

Further volumes of this series can be found on our homepage: springer.com

Vol. 372. António E. Ruano and Annamária R. Várkonyi-Kóczy (Eds.)
New Advances in Intelligent Signal Processing, 2011
ISBN 978-3-642-11738-1

Vol. 373. Oleg Okun, Giorgio Valentini, and Matteo Re (Eds.)
Ensembles in Machine Learning Applications, 2011
ISBN 978-3-642-22909-1

Vol. 374. Dimitri Plemenos and Georgios Miaoulis (Eds.)
Intelligent Computer Graphics 2011, 2011
ISBN 978-3-642-22906-0

Vol. 375. Marenglen Biba and Fatos Xhafa (Eds.)
Learning Structure and Schemas from Documents, 2011
ISBN 978-3-642-22912-1

Vol. 376. Toyohide Watanabe and Lakhmi C. Jain (Eds.)
Innovations in Intelligent Machines – 2, 2011
ISBN 978-3-642-23189-6

Vol. 377. Roger Lee (Ed.)
Software Engineering Research, Management and Applications 2011, 2011
ISBN 978-3-642-23201-5

Vol. 378. János Fodor, Ryszard Klempous, and Carmen Paz Suárez Araujo (Eds.)
Recent Advances in Intelligent Engineering Systems, 2011
ISBN 978-3-642-23228-2

Vol. 379. Ferrante Neri, Carlos Cotta, and Pablo Moscato (Eds.)
Handbook of Memetic Algorithms, 2011
ISBN 978-3-642-23246-6

Vol. 380. Anthony Brabazon, Michael O'Neill, and Dietmar Maringer (Eds.)
Natural Computing in Computational Finance, 2011
ISBN 978-3-642-23335-7

Vol. 381. Radosław Katarzyniak, Tzu-Fu Chiu, Chao-Fu Hong, and Ngoc Thanh Nguyen (Eds.)
Semantic Methods for Knowledge Management and Communication, 2011
ISBN 978-3-642-23417-0

Vol. 382. F.M.T. Brazier, Kees Nieuwenhuis, Gregor Pavlin, Martijn Warnier, and Costin Badica (Eds.)
Intelligent Distributed Computing V, 2011
ISBN 978-3-642-24012-6

Vol. 383. Takayuki Ito, Minjie Zhang, Valentin Robu, Shaheen Fatima, and Tokuro Matsuo (Eds.)
New Trends in Agent-Based Complex Automated Negotiations, 2012
ISBN 978-3-642-24695-1

Vol. 384. Daphna Weinshall, Jörn Anemüller, and Luc van Gool (Eds.)
Detection and Identification of Rare Audiovisual Cues, 2012
ISBN 978-3-642-24033-1

Vol. 385. Alex Graves
Supervised Sequence Labelling with Recurrent Neural Networks, 2012
ISBN 978-3-642-24796-5

Vol. 386. Marek R. Ogiela and Lakhmi C. Jain (Eds.)
Computational Intelligence Paradigms in Advanced Pattern Classification, 2012
ISBN 978-3-642-24048-5

Vol. 387. David Alejandro Pelta, Natalio Krasnogor, Dan Dumitrescu, Camelia Chira, and Rodica Lung (Eds.)
Nature Inspired Cooperative Strategies for Optimization (NICSO 2011), 2011
ISBN 978-3-642-24093-5

Vol. 388. Tiansi Dong
Recognizing Variable Environments, 2012
ISBN 978-3-642-24057-7

Vol. 389. Patricia Melin
Modular Neural Networks and Type-2 Fuzzy Systems for Pattern Recognition, 2012
ISBN 978-3-642-24138-3

Vol. 390. Robert Bembenik, Lukasz Skonieczny, Henryk Rybiński, and Marek Niezgódka (Eds.)
Intelligent Tools for Building a Scientific Information Platform, 2012
ISBN 978-3-642-24808-5

Vol. 391. Herwig Unger, Kyandoghere Kyamaky, and Janusz Kacprzyk (Eds.)
Autonomous Systems: Developments and Trends, 2012
ISBN 978-3-642-24805-4

Vol. 392. Narendra Chauhan, Machavaram Kartikeyan, and Ankush Mittal
Soft Computing Methods for Microwave and Millimeter-Wave Design Problems, 2012
ISBN 978-3-642-25562-5

Vol. 393. Hung T. Nguyen, Vladik Kreinovich, Berlin Wu, and Gang Xiang
Computing Statistics under Interval and Fuzzy Uncertainty, 2012
ISBN 978-3-642-24904-4

Vol. 394. David A. Elizondo, Agusti Solanas, and Antoni Martínez-Ballesté (Eds.)
Computational Intelligence for Privacy and Security, 2012
ISBN 978-3-642-25236-5

David A. Elizondo, Agusti Solanas,
and Antoni Martínez-Ballesté (Eds.)

Computational Intelligence for Privacy and Security

Springer

Editors

Dr. David A. Elizondo
De Montfort University
School of Technology
The Gateway
Leicester
LE1 9BH
UK
E-mail: elizondo@dmu.ac.uk

Dr. Agusti Solanas
Universitat Rovira i Virgili
Department of Computer Engineering
and Mathematics
Av. Paisos Catalans 26
43007 Tarragona, Catalonia,
Spain
E-mail: agusti.solanas@urv.cat

Dr. Antoni Martínez-Ballesté
Universitat Rovira i Virgili
Department of Computer Engineering
and Mathematics Av. Paisos Catalans 26
43007 Tarragona, Catalonia,
Spain
E-mail: antoni.martinez@urv.cat

ISBN 978-3-642-25236-5 e-ISBN 978-3-642-25237-2

DOI 10.1007/978-3-642-25237-2

Studies in Computational Intelligence ISSN 1860-949X

Library of Congress Control Number: 2011940963

Typeset by Scientific Publishing Services Pvt. Ltd., Chennai, India.

Printed on acid-free paper

9 8 7 6 5 4 3 2 1

springer.com

Preface

This book offers a collection of invited papers that present various ways of applying results from the Computational Intelligence domain to privacy and security problems. These papers have been selected among the ones presented at a special session held during the 2010 IEEE World Congress on Computational Intelligence (IEEE WCCI 2010) in July 2010 in Barcelona, Spain.

Malicious actors are very creative when it comes to circumvent mechanisms, tools, techniques and procedures that are deployed to prevent their mischiefs or, at least, to detect them in due time. Similarly, we should be as creative in the way we are looking at new solutions. An interdisciplinary approach to security can only favour cross-fertilization of ideas. This book, by being devoted to novel techniques and developments on the use of Computational Intelligence techniques such as evolutionary algorithms, neural networks, fuzzy systems, etc., is certainly one step into that direction. Similar initiatives should be encouraged to avoid the development of isolated school of thoughts not aware of interesting results obtained by others.

The reader will find in this book results obtained by some of the leading investigators and research groups in this growing area. It will provide him with an overview of the most recent advances in the Computational Intelligence techniques being developed and used to protect the privacy and security of the users of information and communication technologies.

Marc Dacier,
Senior Director,
Collaborative Advanced Research Department
Symantec Research Labs
September 2011

Contents

Computational Intelligence for Privacy and Security: Introduction

David A. Elizondo, Agusti Solanas, and Antoni Martínez-Ballesté

Abstract. The field of Computational Intelligence relates to the development of biologically inspired computational algorithms. The field includes three main areas: neural networks, genetic algorithms and fuzzy systems. This book presents recent research on the application of Computational Intelligence models, algorithms and technologies to the areas of Privacy and Security. These areas are of vital importance to the safety, prosperity and future development of the world's economy, underpin trust in all areas of commerce, defence, security and good governance, and are central to the daily lives of many people.

1 Introduction

The development of the Information Society has entailed an astonishing increase of the number of computer users. In addition, since the mid 90s of the 20th century, Internet has become an essential part of every personal computer, as important as the display or the mouse. Almost everyone in the so-called First World owns or has access to an Internet-enabled computer. The network of networks is currently a network of technologies, a bunch of powerful communication tools and a huge quantity of knowledge: always there, growing and expanding... Waiting for being searched, copied, linked, downloaded, hacked, leaked... Internet is a kind of parallel universe made of bits, and distributed throughout our homes, schools and even our pockets.

Computers and other Information and Communication Technologies (ICT) are not relegated to stay inside a boring, flat, square screen. They are in our skies,

David A. Elizondo
Computer Forensics and Security Group, Department of Computer Technology, De Montfort University, The Gateway, Leicester, LE1 9BH, United Kingdom
e-mail: `elizondo@dmu.ac.uk`

Agusti Solanas · Antoni Martínez-Ballesté
Dept. Computer Engineering and Maths, Universitat Rovira i Virgili, 43007 Tarragona, Catalonia - Spain
e-mail: `{agusti.solanas,antoni.martinez}@urv.cat`

D.A. Elizondo et al. (Eds.): Computational Intelligence for Privacy and Security, SCI 394, pp. 1–4.
springerlink.com

controlling worldwide communications between people, machines, servers and armies. They are in our banks, virtualizing our savings and investments. They are far away from our home, controlling the power generated in nuclear plants. Computers and data networks are the linchpin of our advanced society.

Notwithstanding, there is an evil part present in this modern fairy tale. Since the dawn of the computer era, breaking the security barriers of computer systems has focused the attention of cyber criminals. Some of them only want to play, some others are eager to invade our privacy and steal our money, and the worst try to make nuclear plants explode or the like. Security and privacy are essential for the success of the Information Society. Technology moves fast. Legislation changes slowly. But the knowledge is already there, and growing. In that sense, during the last decades, researchers in computer science and other related disciplines have been developing a plethora of techniques to provide ICT with security and privacy. Although cryptography is one of the first terms that come to mind when thinking about security, there are some other fields that attract the attention of security and privacy experts. Probably, one the most important ones is computational intelligence.

2 Computational Intelligence

Computational Intelligence is based on biologically inspired computational algorithms. The key pillars that compose this field are: neural networks, genetic algorithms and fuzzy systems. Neural networks are algorithms that can be used for function approximation or classification problems [2, 3]. They include supervised, unsupervised and reinforcement learning [4]. Genetic algorithms [1] are search algorithms inspired by biological genetics. They rely on two main operators, cross over and mutation. Populations of individuals representing solutions to the problem are created over several generations. The algorithm uses a random guided approach to optimize problems based on a fitness function. Fuzzy logic [6] is based on fuzzy set theory [5] to deal with reasoning that is fluid or approximate rather than fixed and exact. Fuzzy logic variables have truth values ranging in degree between 0 and 1 which can handle partial truth.

Computational Intelligence techniques have been successfully used in many real world applications on a variety of engineering problems. They can also be used to protect the privacy and security of the users of information and communication technologies. Topics of interest in this ares include: Location Based Services, Statistical Databases, Statistical Disclosure Control, Denial of Service Attacks, Forensics, Intrusion Detection Systems, homeland security, critical infrastructures protection and access control. Two important aspect to consider on the using computational intelligence towards security and privacy are the legal and the ethical ones.

3 Organisation and Plan of the Book

This book presents a compilation of chapters describing novel recent techniques and developments on the use of Computational Intelligence techniques towards the problems of Privacy and Security. A chapter on ethical and legal issues emerging from the use of computational intelligence techniques into privacy and security is also included. The book is divided into fourteen chapters, whose aims are next summarized.

Chapters 2 (Bitter et al.), 3 (Al amro et al.) and 4 (Al amro et al.) give an overview of the three main pillars of computational intelligence: Neural Networks, Genetic Algorithms and Fuzzy Logic respectively and their applications into Privacy and Security.

A topological study of chaotic iterations application to hash functions is presented in Chapter 5 (Guyeux). The authors present a comprehensive study of the topological behavior of chaotic iterations. This tool possesses the property of topological mixing. The authors propose an application example of their methodology for a truly chaotic hash function.

Results on SOM-based Techniques towards hierarchical visualisation of network forensics traffic data are presented in Chapter 6 (Palomo et al.). A novel method for the analysis and visualisation of network forensics traffic data, based on growing hierarchical self-organising maps is presented. The performance of the method is evaluate by clustering traffic data for its visualisation in a hierarchical fashion to enhance the ability of digital forensics to find evidence of attacks or anomalous behaviour in the network. Experimental results show the utility of this approach.

Evolutionary based heuristics for solving soft security issues in MANETs are presented in Chapter 7 (Seredynski et al.). The main idea of this work is to demonstrate how the problem of a selfish behaviour in mobile ad hoc networks can be solved by means of a soft security approach based on evolutionary game theory. The approach combines a classical trust system with an evolutionary game-theoretic model. The authors demonstrate that the numbers of selfish and altruistic nodes have a significant influence on the desired settings of the system.

Genetic algorithms have been shown to be able to find good solutions to the microaggregation problem for small, multivariate data sets. However, they can hardly cope with large, multivariate data sets. In Chapter 8 (Solanas et al.) authors propose partitioning the data into smaller disjoint subsets that the genetic algorithm can handle separately. They elaborate on the study of a partitioning strategy based on the variable-MDAV algorithm, study the effect of several parameters, and compare their approach with the most relevant previous proposals.

Chapter 9 (Sramka) introduces a methodology for breaching data privacy based on data mining techniques. The method proposed by the author employs data mining and fusion to remove some noise from the perturbed sensitive values. Through experimental results the author confirms that his methodology presents a significant privacy risk to published perturbed data because the majority of the noise can be effectively removed.

The use of computational intelligence techniques for feature selection of hand biometrical traits is discussed in Chapter 10 (Luque et al.). The authors propose a novel methodology for using feature selection in hand biometric systems, based on genetic algorithms and mutual information. The aim is to provide a standard features dataset which diminishes the number of features to extract and decreases the complexity of the whole identification process. The experimental results show that it is not always necessary to apply sophisticated and complex classifiers to obtain good accuracy rates.

The construction of a visual surveillance system based on neural networks is discussed in Chapter 11 (García-Rodríguez et al.). The system is based on a variation of self-organising neural networks called growing neural gas. This is modified to cope with temporal constraints. The system is able to track image features in video sequences and it can automatically keep the correspondence of features among frames in the sequence using its own structure.

Recent advances in spam filtering techniques is presented in Chapter 12 (Almeida et al.). E-mail spam continues to have a big economic impact in society. A novel method is proposed based on the Minimum Description Length principle. Results indicate that the proposed filter is fast to construct, incrementally updateable and clearly outperforms the state-of-the-art spam filters.

One of the major problems concerning information security is the detection of malware. Chapter 13 (Shankarpani et al.) presents a method of classifying malicious and specifically concentrates on the performance of kernel methods in the context of robustness and generalization capabilities of malware classification. Authors show that malware analysis based on the API calling sequence and API frequency, which reflect the behaviour of a particular piece of code, gives good accuracy to classify malware.

Ethical and legal issues of raising from the use of computational intelligence techniques in computer security and computer forensics is presented in Chapter 14 (Stahl et al.). A problem that normative discussions of technologies face is that the technical capabilities under investigation tend to be unclear and that the experts in normative questions do not tend to be experts in technical developments and vice versa. The authors study the ethical and legal problems arising from a new and fast moving field, namely that of computational intelligence and its application to computer security and forensics.

The book will be of interest to researchers in industry and academics and to postgraduate students interested in the latest advances and developments in the field of Computational Intelligence for Privacy and Security.

Acknowledgements. The editors wish to thank all the authors who contributed with their research in this volume. They would also like to thank the reviewers for their comments which have enhanced the contents of the chapters. A special thanks goes to Marc Dacier, Director of Symantec Research Labs Europe and US, for contributing with the preface of the book.

References

1. Goldberg, D.: Genetic Algorithms in Search, Optimization, and Machine Learning. Addison-Wesley (1989)
2. Gurney, K.: An Introduction to Neural Networks. Routledge (1997)
3. Haykin, S.: Neural Networks: A Comprehensive Foundation. Prentice-Hall (1999)
4. Masters, T.: Practical Neural Network Recipies in C++. Academic Press (1993)
5. Zadeh, L.A.: Fuzzy sets. Information and Control 8(3), 338–353 (1965)
6. Zadeh, L.A.: Fuzzy algorithms. Information and Control 12(2), 94–102 (1968)

An Introduction to the Use of Neural Networks for Network Intrusion Detection

Christian Bitter, John North, David A. Elizondo, and Tim Watson

Abstract. Modern Society is becoming increasingly dependent upon ever-more complex systems. We are in a situation where a security breach can have an impact on individuals, institutions and critical services, such as power and communication systems. This reliance, along with the possibility of remaining both anonymous and geographically separate from an intrusion, has made cyber-crime an attractive arena for criminals. To protect their assets organisations can use a multi-layered approach to security. As well as the other areas of access control, systems which can detect if malicious or unauthorised activity is occurring are becoming more and more prevalent; intrusion detection systems are at the centre of this. Of particular benefit to intrusion detection systems are any technique with the potential to identify previously unseen patterns, such as neural networks. This chapter is concerned with the state-of-the-art of using neural networks, as part of an intrusion detection system, to identify suspicious or malicious systems traffic. We examine host based systems (where all the information is gathered from a single host) and network based systems. We examine a cross section of different types of neural networks and their application to differing types of intrusion detection.

1 Introduction

Computer security can be considered as a combination of three qualities of an information flow: availability, confidentiality and integrity [1]. For information to be

Christian Bitter
Centre for Computational Intelligence, Department of Informatics, De Montfort University, The Gateway, Leicester, LE1 9BH, United Kingdom
e-mail: christian.bitter@gmail.com

John North · David A. Elizondo · Tim Watson
Computer Forensics and Security Group, Department of Computer Technology, De Montfort University, The Gateway, Leicester, LE1 9BH, United Kingdom
e-mail: {elizondo,tw,johnn}@dmu.ac.uk.

D.A. Elizondo et al. (Eds.): Computational Intelligence for Privacy and Security, SCI 394, pp. 5–24.
springerlink.com

secure its flow must be constrained (security regulations, policies, authorities) appropriately so a user can have confidence in all three qualities. Computer security measures can be applied to protect valuable IT assets (theft, industrial espionage), to detect unauthorised access or misuse, to fulfil legal compliance requirements (threat modelling, audit, break-in analysis), or to prevent data loss in case of disasters. A combination of the ever-expanding nature of communication networks (computer, telephony), an increased number of services using these networks and societies increasing dependence on these services, has raised awareness of the need for computer security. Services like electronic commerce, financial operations, power supply and monitoring are at the heart of modern cultures. The increased use of services by an expanding section of the population has spurred criminals' interest, who want to capitalise on information society's dependency on IT. Cyber-terrorism and cyber-crime have grown significantly in the last few years, partly due to the reasons mentioned already, along with an decreasing entry level of knowledge needed to perform attacks, sinking hardware and operation costs and wide-spread access to communication networks. In addition cyber-crime can be seen as a low risk activity, as a perpetrator can remain a long distance away from an attack, has a reasonable chance of remaining anonymous and due to national differences both in laws and regulations and in their enforcement.

For the remainder of this chapter we focus on intrusion detection. Intrusion detection is a vibrant field of computer security research involving the analyses and reasoning about a digital environment's current state of security, and so enabling external components (software, hardware, human) to take action if required. An intrusion can be defined as a successful set of attempts to compromise a computer or a computer network's security. Intrusion detection is the process of detecting inappropriate, incorrect, or anomalous activity targeted at computing and networking resources, to identify unauthorised use, misuse, or abuse of computer systems to strengthen security and increase resistance to attacks [1–3]. Intrusion detection is implemented via intrusion detection systems (IDS).

Two commonly used methodologies to implementing IDS exist, namely (i) signature-based (also known as pattern based) and (ii) behaviour-based (or anomaly based) detection. Signature-based detection uses signatures of known attacks to detect intrusions via pattern matching on gathered environmental information (log data, events). Signatures represent a defender's domain knowledge which makes them intuitively appealing. Unfortunately, signature-based intrusion detection is not very robust with respect to unknown signatures, including variations of known ones; because of this signatures have to be kept up-to-date at all times to ensure proper detection. Behaviour-based detection profiles user behaviour and classifies it either as being normal or abnormal. A model of normal user activity, with respect to some environment, is used to classify behaviour as normal or not normal and accept or reject that behaviour accordingly. This makes behaviour-based approaches more robust with respect to unknown attacks but may lead to legitimate activity being reported as an intrusion.

Desired characteristics of an IDS may include accuracy, performance and completeness [2]. Accuracy is concerned with maximising the number of correct detections while at the same time minimising the number of false-positives (legitimate action flagged as intrusion) and false-negatives (intrusion flagged as legitimate action). Clearly, the issue of false-negatives is more serious than false-positives, since an organisation's computing resources might be compromised unknowingly as opposed to an administrator receiving a false alarm. Performance includes measures such as intrusion detection rate, i.e. throughput of event-processing. Completeness refers to an IDS being able to recognise all known intrusion attacks. Although this is desirable it is not achievable due to the transient nature of attacks and the comparative advantage that attackers have.

The aim of this chapter is to demonstrate that Neural Networks (NNs) are a viable approach to building robust, adaptable and accurate IDS. Selected state-of-the-art IDS are discussed, which employ NN to detect suspicious activity (intrusion, attack). Intrusion prevention systems (IPS) are often considered as a functional extension of IDS. They aim not only at detecting but also blocking and preventing identified malicious traffic and suspicious undesired activity from affecting an IT system via, for example, changing firewall rules [4]. IPS could significantly benefit from leveraging artificial intelligence (AI) techniques such as NNs to perform intrusion detection. IPS will not be covered in this chapter but may, in a future paper.

The rest of this chapter is structured as follows. Section 2 presents a classification of IDS discussed throughout this chapter. Section 3 presents an introduction to intrusion detection using different types of Artificial Neural Network (ANN) technologies. Section 4 Discusses the use of other NN types such as Functional Neural Networks and Unsupervised Neural Networks. Section 5 then discusses other related issues, such as the dataset used for testing and issues affecting the reliability and performance of differing Neural Networks. We finish with our conclusions in section 6. For an introduction to pattern classification and computer-based learning see [5–7].

2 Classification

In this section, a classification is discussed, which arranges the presented approaches to intrusion detection into a logical structure. There is potential for a wide range of classification criteria to choose from, for example, one might classify NN-based IDS according to the learning strategy or topological properties. Although these criteria allow the presented systems to be arranged into a coherent framework, such a framework would not be very informative, because a NN's topological properties can be specific to the particular knowledge used and might have been obtained using a reverse-fitting approach, i.e. choosing a topological parameter value, testing its effect on performance and making modifications accordingly. This process would be complicated further as we examine different varieties of Neural Networks. Instead we chose five classification criteria, four of which originate from computer security and intrusion detection (location, time, scenario, strategy) and one originating from the learning strategy domain (model). The resulting classification is presented in Table 1.

Location refers to where an IDS collects information to analyse from. This gathering of information may be either direct using sensors or indirect using a supplied data set. IDS collect and analyse information originating from a single host (Location=host), from a network (Location=distributed) or hybrid, like agent-based IDS (agents migrate between network nodes). Some authors distinguish even further between application-based and vulnerability-based [2, 8] IDS; this notion is not assumed here, since application-based IDS can be regarded as a specific subset of host based IDS and vulnerability-based IDS can be regarded as a specialisation of distributed IDS. The model category captures an IDS's classification mechanism, such as ANNs, generalisations of ANNs like FNN or variations such as Self-Organising Map (SOM).

Time refers to the timeliness of an IDS's learning/training operation. Time may either be on-line or off-line. On-line training, also referred to as incremental or sequential training, is able to perform the learning operation piece-by-piece in a serial fashion, i.e. on one individually (randomly) selected training sample at a time. Consequently, on-line learning can be conducted without having the entire training data available from the beginning or a potentially infinite training data set. In contrast, an approach classified as Time=off-line uses off-line or batch mode learning. This learning takes place on the whole problem data, i.e. all training data is presented in one learning iteration/epoch to the learning algorithm. The learning algorithm tries to adapt to the whole learning data set and the next iteration of learning is performed until some termination criterion (performance, time) is reached. A comparison of ANNs on-line and off-line training with respect to automobile industry and bridge behaviour under seismic activity can be found in [9]. The authors mention that on-line training is more likely to find the best solution (best model) and is resistant to redundant training data, i.e. not negatively affected by multiple presentations of the same training sample. Please note that any system trained on batches of the DARPA KDD Cup 99 (DK99C) data set [2], a gold standard data set, is classified as Time=off-line and Location=distributed. The reason for this is that the data set includes a wide variety of intrusions simulated in a military network environment (distributed) and training is not performed incrementally. DK99C data set consists of (i) two million unlabeled test data connection records (two weeks of TCP dump data), and (ii) five million labeled training data connection records (seven weeks of network traffic compressed binary TCP dump data). Each training connection record is labeled as normal or as attack, with exactly one specific attack type of 22 different attack types. Each of the 22 attack types can be categorised into one of four main categories: denial-of-service (DoS), unauthorized access from remote machine, unauthorized access via local root privileges, and surveillance and other probing. Each connection record has 41 qualitative and quantitative features. Around 20% of the data is normal (no attack). Besides the DK99C data set, other data sets for constructing and evaluating IDS exist. One such data set is the DARPA 98 data set resulting from DARPA's first sponsored intrusion detection evaluation effort in 1998. In 1999 and 2000 further evaluations were carried out by DARPA leading to the DARPA 99 and 2000 data sets respectively. These data sets extend the DARPA 98 data set in terms of novelty and complexity of attack, i.e. unknown types and attacks comprised of multiple distinct steps. Quite recently, NSL-KDD data set was proposed to circumvent limitations experienced with the DK99C data set, such as redundant and duplicate records in the training and test data set [10]. Although these publicly available standard data sets exist, researchers sometimes use custom data, for example data obtained from log files of a departmental or privately run web server to evaluate their approach.

The strategy category item refers to the two strategies IDS employ to detect intrusions, i.e. behaviour-based or signature-based intrusion detection. If a system makes use of both strategies it is categorised as a hybrid system. The final category, scenario, refers to the particular domain the respective IDS system is applied to, i.e. unspecified (IDS) or specific intrusions like DoS, spam or worm attack.

Table 1 Classification of discussed IDS

System/ Author	Citation	Location	Model	Time	Strategy	Scenario
Chang et. al	[11]	distributed	ANN	off-line	behaviour-based	IDS
Wu	[12]	distributed	hybrid	off-line	hybrid	SPAM
IMA-IDS	[3]	hybrid/agent	ANN	off-line	behaviour-based	IDS
MOVICAB-IDS	[13]	hybrid/agent	hybrid	on-line	hybrid	IDS
Alonso-Betanzos et. al	[14]	distributed	FNN	off-line	behaviour-based	IDS
DoSID	[2]	distributed	ANN	off-line	behaviour-based	DOS
Iftikhar et. al	[8]	distributed	ANN	off-line	behaviour-based	DOS
Stopel et. al	[15]	host	ANN	off-line	behaviour-based	WORM
MPCA-MSOM	[16]	distributed	SOM	off-line	behaviour-based	IDS
ESOM	[17]	distributed	SOM	off-line	behaviour-based	DOS

3 Selected Applications of ANN and Related Approaches to IDS

An *Artificial neural network* (ANN) is a biologically-inspired computation device to simulate structural and functional aspects of the human brain. Clearly, for there is a great deal of complexity inside biological nervous systems which have not even been fully understood to date, ANNs are only an abstraction.

ANNs consist of a group of non-linear highly interconnected processing elements - artificial neurons. An artificial neuron converts some weighted input vector via an activation function (weighted sum) to some desired output vector by passing the result of the activation function to a transfer function (e.g., nonlinear sigmoid transfer function). All ANNs have a set (layer) of neurons called input neurons, receiving inputs from the outside world (sensor data). Input neurons belong to the input layer of an ANN. Neurons passing an ANN's final result to the outside world are called output neurons and belong to the output layer of the ANN. If an ANN has further neurons not belonging to an input or output layer, the ANN is said to have internal or hidden neurons which belong to the ANN's hidden layer. Hidden neurons receive inputs from other neurons or themselves (feedback). In case an ANN has no feedback, it is said to be a feed forward (FF) ANN, consisting only of neurons which receive their input from neurons of a previous layer and send their output to neurons of the next layer. FF ANNs are static with respect to the current inputs. On the contrary, if there is feedback (cycle) within an ANN, the network is a recurrent ANN. Consequently, output of recurrent ANN depends on past inputs, i.e. is dynamic. Weights associated with neurons' interconnections are determined during a training period, i.e. an ANN is presented with input it has to mimic best with respect to some expected output and an evaluation scheme (learning strategy) of what best means. By modifying connections between artificial neurons, the ANN is able to adapt to expected outputs - learning.

ANNs have successfully been applied to a broad range of problems, such as medicine (prediction of heart disease), finance (credit assignment), and engineering (machine monitoring). ANNs are an ideal candidate when the need for prediction, classification or control in dynamic and complex environments arises. The reason for this can be found in the following ANN-innate advantages [2, 5, 18]. Firstly, ANNs process information in parallel (**Parallelism**). Secondly, ANNs learn from knowledge implicitly encoded in example data (**Learning by Example**), so a user must only know about the phenomenon to model and how it is represented in the data. Thirdly, ANNs linearly combine several relatively simple nonlinear processing functions to handle complex nonlinear functions, such as error signals in industrial machinery (**Nonlinearity**). Consequently, ANNs are superior to complex, oftentimes inaccurate and incomprehensible linear modeling tools like differential equations. Fourthly, ANNs are capable of working with noisy and incomplete data (**Resilience to noise and uncertainty**). ANNs are able to generalise from learnt observations and process unknown observations or variations of already learnt observations. Fifthly, ANNs are not limited to a particular learning paradigm (**Versatility and flexibility**). Whenever supervision is not feasible (cost) or possible (legal requirements), a particular problem can be modeled using ANNs and unsupervised learning. Lastly, ANNs are an abstraction of biological neural cells making them intuitively appealing (**Intuitive**).

In the following subsection, ANN-based approaches to intrusion detection including DoS, worm and spam detection are presented (see Subsection 3.1 to 3.3). The use of a Functional Neural Networks are discussed in 4.1. IDS using a variation of the NN paradigm, self-organising map (SOM), are presented in Subsection 4.2. In Table 2 the presented IDS' reported accuracy (0%-100%, n.a. were not reported), i.e. the IDS' ability to correctly detect a particular type of anomalous system state, is shown. Approaches to intrusion detection employing Biologically-inspired Complex Adaptive Systems such as artificial immune systems, genetics-based machine learning, swarm intelligence or co-evolution are discussed by Shafi and Abbass [19]. A survey presented by Karim discusses intrusion detection from a computational intelligence standpoint touching on IDS implementations using soft computing or agent-based systems [20].

Table 2 Reported Accuracy values of discussed IDS

System/ Author	Accuracy
Chang et. al	n.a.
Wu	99.6%
IMA-IDS	n.a.
MOVICAB-IDS	n.a.
Alonso-Betanzos et. al	92.45%-99.83%; AVG=96.24%
DoSID	58%-100%; AVG=83.14%
Iftikhar et. al	79%-100%; AVG=96%
Stopel et. al	99.96%-99.97%; AVG=99.97%
MPCA-MSOM	97%
ESOM	98.3%-99.81%; AVG=99.06%

3.1 Selected ANN-Based Approaches to Intrusion Detection

As mentioned before, ANNs have been applied to detect potentially unauthorised actions within today's complex and dynamic computer environments. For example, Barika et. al extended an existing agent-based IDS called IMA-IDS to include an ANN-based decision layer [3]. The motivation to use an agent-based approach is to avoid a single point of failure in the monitored environment. This makes it more difficult for an intruder to learn about the IDS' decisions. Load is balanced across multiple network sites and in case one IDS site fails (DoS attack), all other IDS' sites stay operational. If the network expands in number of nodes, IMA-IDS may scale accordingly via cloning of agents. Updating IDS agents may be done asynchronously, without down-time of the whole system, relying on agent migration and intercommunication.

IMA-IDS is comprised of a four layer architecture implemented in Java. One or more sniffer agents (down level) are responsible for patrolling the network sites and collecting event information from them. Sniffer agents may wander from site to site and clone themselves on-demand. A filter agent (pretreatment level) aggregates and merges event information with respect to source, protocol, time and establishes a baseline of normal system behaviour. At the kernel level, Filter agent analyses processed information and prepares it for the upper level. Decision agent (upper level), uses ANNs to decide whether the environment is currently in a normal or abnormal state (behaviour-based intrusion detection). IMA-IDS' ANN is a FF ANN with 5, 10, and 15 neurons in each hidden layer. Decision agent's ANN was learnt under supervision using a 25-dimensional training data set.

The mentioned approach can be seen as the ANN analogy at the level of processing, i.e. individual agents corresponding to artificial neurons communicating with each other to form global from local knowledge. It would be interesting to apply learning at the agent level. Having a distributed IDS accommodates for the distributed nature of communication networks and attacks launched to compromise them, i.e. no single point of failure, scalability, redundancy to achieve resilience and reliability. Unfortunately, the authors did neither evaluate their approach with benchmark data like the DK99C data set nor mention and discuss issues, such as "How to integrate data from disparate and heterogeneous sites?", "How to ensure a secure agent platform?" or "What happens if an agent becomes malicious?".

Herrero et. al present an agent-based approach to intrusion detection utilising soft computing techniques called Mobile Visualization Connectionist Agent-Based IDS (MOVICAB-IDS) [13]. MOVICAB-IDS combines explicit knowledge using case-based reasoning (CBR) with implicit knowledge via NN-based classification. It is built from six types of agent, namely Sniffer, Preprocessor, Analyzer, ConfigurationManager, Coordinator and Visualizer.

Sniffer functions as a network sensor collecting and segmenting network traffic data. Sniffer passes collected data to Preprocessor agent. Preprocessor, ideally closely located to Sniffer, implements data preprocessing (data cleaning, feature selection). Analyser agent, is an adaptive intrusion classification component combining mentioned intrusion detection techniques (CBR, NN). Data records are processed by Analyzer following the retrieval-reuse-revision-retention CBR process. Firstly, Analyzer determines the most similar (euclidean distance) case in its case base for some collected data point. The retrieved case contains problem and solution features. Solution features are parameters of the NN learning model (learning rate, etc.) used to generate a training session set whose size increases/ decreases in response to dis/similarity between collected data point and retrieved case. The training session set is built from varying model parameter configurations. Then the NN is trained with respect to generated training sessions. Afterwards, the user is presented with NN's training performance. She has to visually select the training session instance that captures best the current network state using Visualizer. Afterwards the case base is updated (collected data point, user selected training session instance). ConfigurationManager is a centrally queryable configuration parameter store. Coordinator agent functions as an agent request-task scheduler.

Herrero and Corchado provide recommendations on how to segment network traffic data to optimally train NN component [21]. They propose simple and accumulated traffic segments. Simple segments encompass traffic originating from a specific time window including overlap time to capture events spanning multiple simple segments. Accumulated segments span multiple consecutive simple segments (without segment overlap) to create an evolutionary view of network state. MOVICAB-IDS is an attempt to build a possibly distributed learning and thereby adaptable IDS. Adaptability can help to increase reactivity in case of varying network load and to correctly classify previously unseen instances of network state. MOVICAB-IDS' agent-based architecture ensures scalability. Cloning of agents helps to increase fault-tolerance. Mobility of agents enables to reduce the IDS' communicational footprint via local instead of global agent communication. Combining visual data exploration (Visualizer) and automated intrusion detection (Analyzer) acknowledges (i) human's cognitive and creative abilities, and (ii) computer's raw computing capabilities in order to analyse large dynamic volumes of data for intrusion detection.

Unfortunately, the authors did not evaluate MOVICAB-IDS using a standard data set and standard intrusion detection measures (accuracy, etc.). For this, validity of the proposed approach and traffic segmentation guidelines cannot be assessed. Although recommendations on traffic segmentation are intuitive, used segmentation parameters (time window length) are not justified and seem arbitrary. Evaluation data set descriptive statistics (dimensionality, distribution) are not presented. Topological properties of the NN Analyzer component are not discussed. Although the title mentions mobile visualization, the authors do not discuss ramifications or benefits of targeting intrusion detection visualisation for mobile devices. Visual analysis of port sweep attacks performed as evaluation is nowhere substantiated, i.e. patterns used as indication for attacks are not explained, test data generation is not described and features used for visualisation are not mentioned.

In large interconnected computer networks enormous amounts of high-dimensional sensor data might be accumulated. Increasing learning data (amount, dimensionality), increases training time needed by an ANN-based IDS. Also, expanding training observations (number, dimensionality), increases the ANN's risk of failing to converge to a stable state. For this, Chang et. al propose an ANN-based IDS which incrementally increases the amount of training data [11]. Their underlying assumption is that normal behaviour is the predominant state of a distributed systems and consequently, will be the dominant observation among training data records. The authors identified two methods to reduce ANNs training time, (i) sample-query and (ii) attribute-query. Sample-query means that the ANN is learnt on an initially small training set. This training set is incrementally enlarged using observations specifically chosen by an oracle component. Selected observations, named sample-query, are representative of specific intrusion attacks, i.e. are far on either side of the decision boundary (close to ± 1). For this, oracle-picked observations improve the learning performance, in contrast to observations which have low information value, i.e. are close to the decision boundary. Attribute-query denotes feature subset selection, i.e. the removal of features from training data. This is achieved using information gain to measure the contribution of a feature to an observation's classification result. Features are removed iteratively, i.e. a feature is tested and removed, if it does not contribute significantly to the observation's classification result. Consequently, using sample- and attribute-query, an ANN is trained on a small yet highly representative portion of the intrusion detection problem data set. Starting from a random initialisation of the ANN's weights, the ANN is trained until the training performance goal Mean Square Error (MSE) = 0.01 or maximum number of training epochs is reached. Training is performed on a subset of the DK99C data set using stratified random sampling, incrementally increasing the initial strata using oracle-chosen sample-query.

Chang et. al evaluated their approach against an equivalent back propagation ANN (BPNN) without sample and attribute-query. The authors report that ANN training time could be reduced by an order of magnitude. Intrusion detection performance standard measures (detection rate, false-positive rate) are very similar to the compared BPNN, with a notable positive exception being the better performance against unknown intrusions. This shows that the authors' approach has good generalisation capabilities. Chang et. al show that data in a supervised learning environment should be apt to the learning task. Conversely, the often mentioned assumption more data is better data, does not always hold since, model-building might take too long or fail at all. Unfortunately, the authors do neither present an evaluation of different attribute subset selection measures nor results of their attribute subset selection approach. Also their motivation to use a particular BPNN topology neither is nowhere motivated nor a comparison of alternative ANN topologies is presented.

3.2 Detection of Denial-of-Service Attacks Using ANNs

A specific instance of unauthorised intrusion attack is the DoS attack attempting to attack a host, network or specific application to make its offered services unavailable. DoS attacks can be distinguished into (i) DoS by saturation, which involves flooding a system's service request queue rendering it unable to respond to genuine requests, or (ii) DoS by vulnerability exploitation targeting flaws in a machine's system stack (OS, application).

Iftikhar et. al demonstrate the applicability of FF ANNs to detection of DoS attacks following a behaviour-based approach [8]. Their ANN has two hidden layers with 14 and 9 neurons each. The input layer has 41 neurons corresponding to the 41 features of the DK99C data set. Two output neurons realise the classification of DoS attacks. The authors designed and implemented their approach in Java, following a five-step process model: data set training/ testing (supply of training data), preprocessing data set (data cleaning, conversion, normalisation), determining the ANN (topology), training the system and testing the system. The ANN was trained under supervision, i.e. labeled observations from the training set are either classified as DoS attack (abnormal) or normal environment state. The authors used resilient BP learning, i.e. only the sign not magnitude of the gradient is used to update neurons' weights and bias [22]. This leads to faster model convergence and ANN's ability to escape local minimum.

Training was performed on a subset of the DK99C data set, using MSE as training performance measure and a maximum training time of 1000 epochs. After model-building, the ANN was tested on an unknown subset of the DK99C data set. Results show that the presented approach performs good with respect to DoS detection, DoS attack classification and overall false-positive and false-negative rate. Detection rate of DoS categories is mostly above 99%, with the exception being the Teardrop DoS attack, which was detected only in 79% of the cases. False-positive and False-negative rate was also close to 0% in nearly all cases, again, the exception being the Teardown attack, having 15%/6% false-positive/false-negative rate. Iftikhar et. al show how resilient BP learning can be applied to DoS attack detection. Advantages of resilient BP include, i.e. decreased likelihood of failing to converge during training and allowing to overcome local minimum. Although mentioned the authors did not consider evaluating different artificial neuron activation functions. Unfortunately, the presented approach was not evaluated on a test data set of appropriate size. For this, results although true, must be taken carefully.

Alfantookh developed a DoS detection system, named DoSID, relying on FF ANNs as well [2]. In contrast to Iftikhar et. al, he runs a series of experiments to choose the best ANN from different configuration of topological properties, such as number of hidden layer neurons 24/64, training epochs 1000/5000, weights initialised to all 0,random or initialised from a trained ANN. The training performance goal is minimising MSE. Model training and testing was performed on the DK99C data set. Evaluation of individual models led to choosing the FF ANN with 24 neurons in the hidden layer and weights initialised to weights of an ANN trained for 200000 epochs. The resultant ANN has one input, hidden and output layer.

To increase DoS classification performance, Alfantookh proposes two improvements, namely gray area and distribution. Gray area increases confidence in a classification result whether it be attack or no attack, because values close on either side of the decision boundary are ignored as being unclassified. This is achieved by extending the zero-width decision boundary to a decision boundary of user-defined width, i.e. $[-1 \leq x_1 \leq 0 \leq x_2 \leq +1]$. Consequently, confidence in classification results is increased and depending on how the gray area is configured, false-positive and false-negative rate may be reduced. The distribution improvement aims to answer the question of how a reasonable gray area can be obtained. For this, ANN training output area $[-1, +1]$ may be divided into fragments of equal width. Each fragment is assigned the number of training observations having output of the fragment's extent (e.g., left and right interval boundary). Fragments having low assignment rates, i.e. are infrequently output as part of the ANN's training, will be regarded as gray area.

The chosen ANN was evaluated without gray area and with a gray area extending from $[-0.8, +0.8)$. Evaluation has shown that gray area improves classification of FF ANNs with respect to false-positive (decrease from 8.57% to 5.71%) and false-negative rate (decrease from 40% to 8%) thereby increasing overall classification accuracy. Alfantookh's approach has provided contributions in the following areas: (i) gray area and (ii) distribution show how classification performance can be increased while false-negative and false-positive rate can be reduced significantly. Decreasing false-negative rate is extremely critical, since a successful DoS attack might mean a disruption in the ability to serviceability of clients. Also, minimising false-positive rate is important with respect to administrators needing trust and confidence in an IDS' classification results. Alfantookh has also emphasised the importance of appropriately initialising ANN's neural weights to reach training convergence.

3.3 Application of ANNs to Spam and Worm Detection

Not always may an intrusion or an attack against hosts be committed directly by a human but by malicious software (malware). A single host within a distributed system infected by malware, can lead to loss, or unauthorised utilisation/ modification of sensitive data. Also, abrupt and forced termination of provided services can cause users to question the current state of the computer environment. A computer worm is executable code in the form of a binary image, code injected into a running thread or application-interpretable source code. It is able to self-replicate over a communication network and does not need user interaction to do so. Computer worms do not always contain harmful routines but may cause damage simply by binding resources (CPU, network).

Stopel et. al propose ANN trained under supervision to malware identification based on behaviour classification (normal, abnormal) for the Microsoft Windows platform [15]. The authors infected several computers with seven real worms to generate training and test data. Over 60 features like processor or operating system state were gathered from infected (abnormal behaviour) and non-infected (normal behaviour, "clean" worm) machines. Performing two separate test runs, data was collected from infected and non-infected machines with and without user activity performed on them. After collection, test data was cleaned and processed to ensure that all worms are correctly reflected in recorded measurements. The author tested worm classification using an ensemble of learning techniques like ANNs, decision tree (DT), k-Nearest-Neighbour (KNN) and ANNs clustering analysis (ANNCA). ANNCA uses that hidden layer neurons of well-trained ANNs have output values very close to 0 or 1 (sigmoid transfer), i.e. are far from the decision boundary. If rounded down/up, hidden layer neurons' output vector may be interpreted as a binary pattern vector, which corresponds to a cluster description. Cluster descriptions may be arranged in a cluster matrix, such that rows correspond to clusters and columns to classes (specific worm, clean).

To learn about the relative importance of each of the selected 68 features, a feature ranking analysis was conducted. Ranking all features led to the identification of top-5 classification features. The authors evaluated all approaches against each other with respect to standard measures like accuracy, or false-positive rate. Results show that in case of no user activity being performed on infected computers, ANN outperforms DT and KNN. However, when tested against machines with user activity, DT performed best. Also, ANN's ability to handle unseen observations, i.e. generalise from learnt data, is advantageous when the need to classify new or variations of existing worms arises. Evaluation of ANNCA led to mixed results, i.e. some worms were classified better and some were classified worse than by standard ANN-based classification. Also, in the start of a worm's lifetime ANN classification performed worse than for older worms, which can be attributed to a worm initially not revealing itself as such.

Stopel et. al show that worm detection is possible using operational parameters (CPU usage, memory usage) of the monitored computer system and ANN-based classifiers. ANNCA has demonstrated how domain knowledge (worm detection) can be extracted from ANNs and subsequently be transformed into a different representation, which might reveal further insight into the problem to solve. Conducted experiments led to the creation of data set, which could be enhanced and supplied as a reference data set. Open issues are: (i) distributed nature of worm attacks, i.e. monitoring several systems and correlating gathered measurements to perform worm detection, (ii) reducing ANNs training time, (iii) identifying and ranking more features useful for worm classification, (iv) incorporating the hiding nature of worms and (v) evaluation against other worm detection approaches.

Increased reliance on email as a means for private and corporate communication, has also increased fraudulent activity performed using massively sent email to target an enormous audience at almost no cost. Email sent to arbitrary recipients without their consent (intrusion) for the sender's disproportionate benefit, is called spam. Spam is sent by spammers to sell products or services to potential customers. Spammers masquerade spam using a variation of techniques like delivery during non-office hours (larger bandwidth available, lower risk to be identified), hide origin of sender and message's route, using interesting, welcoming or stimulating subject line. Automatically separating spam from non-spam (ham) is an equally important and challenging task. Commonly used approaches to spam detection are blacklisting (flagging illegitimate senders), machine learning (ML) (classifying ham and spam) and keyword filtering.

Wu presents a spam detection method using ANNs and rule-based techniques [12]. The author regards spam as the result of spamming behaviour, i.e. behaviour of spammer manifests itself in properties of spam. Because of good reported performance in the domain of spam detection, Wu combined ANNs- with a rule-based approach to detect spam in Simple Mail Transfer Protocol (SMTP) delivered mail. He identified several spam classification relevant features from email header (h-features) and Mail Transfer Agent (MTA) syslogs (s-features). Feature subset selection (b-features) led to the reduction of feature space thereby reducing complexity of the overall task. b-features were combined pairwise leading to x-features. Wu models spamming feature as a five-tuple composed of header, syslog, and x-features. b- and x- features are used to train a BPNN in the two-staged spam classification process. The first stage is comprised of the rule-based email preprocessing step, which uses mentioned email features and hand-crafted if-then rules, stored in a rule database, to determine whether some email exhibits a particular spamming behaviour or not. For example, rules evaluate whether or not the sender address appears to be legitimate. The second stage makes use of a BPNN to classify preprocessed emails. For this, b- and x-features are encoded to be used for ANNs training and testing. Training and test data was collected from MTA. Training data was labeled using golden rules (domain expertise). The resulting BPNN has one input layer with 72 input neurons (b- and x-features), two hidden layer with 10/6 neurons and one output layer neuron classifying an email as being spam or not. The first hidden layer collects b- and x- features separately, whereas the second hidden layer combines all features. The ANN was learnt using gradient descent learning and minimising root MSE (RMSE) as the training performance goal. Wu evaluates the proposed approach against a keyword filtering and various ML-based systems, using standard measures including misclassification rate, and accuracy. Results indicate that behaviour-based spam detection leads to better classification results than (i) pure evaluation of header and syslog features, (ii) keyword filtering approaches due to the transient nature of keywords used by spammers and (iii) compared ML systems. However, it is unclear whether this can be attributed to Wu's approach or the chosen ML systems. Wu shows that spam can be recognised as the result of spam-behaviour, which might change infrequently over time, as opposed to keywords used by spammers. Consequently, spam detection can be conceived and approached using rule-based and ANNs techniques capturing explicit/implicit knowledge present in spammer's behaviour and spam email features.

4 Variations upon the Neural Network

There are numerous different variations of neural networks. So far we have concentrated upon the most popular, the artificial neural network. While discussing the different ways the ANN has been applied to intrusion detection we have touched on different ways it can be implemented, such as feed forward networks and back propagation networks. In this section we will look at two completely different types of neural network, the functional neural network and the self organising map. We will look at how they are different and their uses in intrusion detection.

4.1 Variations of Neural Network – Functional Neural Network

A generalisation to ANN, called functional NN (FNN), is proposed by Alonso-Betanzos et. al [14]. FNN differ mainly from ANNs in that FNN weights are incorporated into the artificial neuron's transfer function as function parameters. Consequently, the transfer function, from a given family of functions (polynomial, Fourier), is unknown in its coefficients, which have to be learnt in the FNN's training phase. For example, a polynomial neuron function might be $f(x_1) = \sum_{i=1}^{m_i} c_i x_1^i$, where coefficients c_i would have to be trained leading to different functional neurons. FNN both incorporates knowledge about the domain to be modeled and the peculiarities about the data used to model the particular problem. Domain specific knowledge is captured via the FNN's topology, whereas knowledge about the data is reflected by the structure of individual functional neurons. Functional neuron output is provided by an associative output function node, aggregating the input vector. The authors compared polynomial, Fourier and exponential FNN with each other and found that given the DK99C data set, a FNN having five functional nodes with three unknown coefficients each delivered the best results (error rate, detection rate, false-negative rate). A direct evaluation against the DK99C winner led to the finding of FNN having lower error rate and good generalisation capability, i.e. good performance against unknown intrusion patterns. However, FNN seem to show a higher false-positive rate than the DK99C winner but lower when compared with support vector machine (SVM) or linear perceptron.

Alonso-Betanzos et. al show that despite their long presence, ANNs can be further extended and enhanced. Whether or not FNN's performance is fix or can be transfered to another domain must be evaluated. Also, different function families and their performance must be analysed. A claim that FNN allow for incorporating data interdependencies into the classifier is nowhere substantiated. Lastly, further research on FNN structural and operational properties, and application scenarios (including comparative case studies with ANNs) must be conducted to possibly provide a true generalisation of ANNs.

4.2 Unsupervised Neural Network – Self Organising Map

From the many variations of NN, Kohonen's self-organizing map (SOM) is one of the most popular NN models [7, 23]. Like ANNs, SOM have their origin in biology. SOM can be categorised as unsupervised or competitive learning. SOM is unsupervised, because there is no target vectors to be learnt. SOM is trained until convergence, i.e. the SOM's neurons, often arranged in a two-dimensional rectangular grid, form a discrete topological mapping of the input space – self-organise (no external influence) according to presented observation. Reaching convergence depends on size and dimensionality of training data. SOM is competitive, because there is only one winning neuron in the output layer. Trained SOMs create clusters of data, where similar feature vectors are located in a specific region in the output space, i.e. SOM can provide topologically preserved mapping from input to output space, which enables for discovering clusters and relationships among data points. This makes SOM resilient with respect to faulty, noisy or unknown observations. The topology-preserving mapping from input to output space is a multi-dimensional scaling method, i.e. a nonlinear dimension reduction from input to output space dimension. For this, SOM has been successfully applied in the fields of vector quantization, dimensionality reduction, data visualization, clustering and classification [18]. SOM offers the following advantages, with respect to intrusion detection: visual representation of environment state, visual presentation of relationships and concept implicitly present in data, requires only little domain knowledge.

Bai et. al propose a combined behaviour-based approach to intrusion detection. Their approach, called MPCA-MSOM, combines principal component analysis (PCA) and hierarchical SOM to apply it to intrusion detection. PCA is used to reduce dimensionality of input space. Hierarchical SOM is used to overcome the SOM limitation of single-class classification, thereby improving classification accuracy. The authors present a new neighbourhood function, which extends reach from one output neuron to several neurons in the primary output neuron's neighbourhood, thereby extending traditional SOM to the mentioned SOM hierarchy. MPCA-MSOM was trained and tested using the DK99C data set. The author evaluated MPCA-MSOM performance against traditional SOM and KNN. Results show that MPCA-MSOM is superior to traditional SOM and KNN with respect to detection and false-positive rate. The presented approach shows that hierarchical SOM may increase classification performance of traditional SOM and PCA reduces dimensionality of input feature vector, leading to reduced model-build time and increased overall performance. MPCA-MSOM demonstrates the applicability of SOM to building IDS.

As noted, SOM has been successfully applied to data visualisation and clustering. Mitrokotsa and Douligeris propose a variation of SOM, called emergent SOM (ESOM) to tackle the DoS detection problem following a behaviour-based approach. ESOM learning is not restricted to particular SOM output neurons but incorporates the complete neuron map [17]. Incorporation of the complete SOM during learning allows to surface higher order concepts otherwise remaining hidden in input data. ESOM has been trained on DK99C data set. A preprocessing step was conducted to perform feature subset selection. Features were selected using domain expertise. Evaluation using standard measures like detection , false-positive and false-negative rate showed that ESOM detection performance is quite stable. A trained ESOM can be transformed to a three-dimensional (2-D and colour) representation of detected DoS attack patterns using U, U* or P-matrix. This transformation to visualise classified attacks presents the true advantage of Bai et. al's approach. Visualisation can be used by operators and system-administrators to not only understand DoS attacks or why certain behaviour is marked as such but also to appropriately react on it. A drawback of the presented approach is the high computational complexity caused by large training data sets.

5 Further Topics

Designing and implementing an ANN-based IDS is not easy because of the complexity of the problem domain and solution approach. For example, choosing the wrong or imbalanced data (feature, resolution, format, age, distribution) might lead to unexpected results, no results or to results which cannot be used to draw conclusions or perform generalisation. Also, issues such as legal compliance or organisational security must often be taken into account when planning on implementing an IDS. For this, it might be of importance to pay attention to some of the issues presented in the following sections.

5.1 Choosing a Data Set

Although often referenced and used as a gold standard, the DK99C data set has been reported as partially inappropriate for modeling an IDS [24]. Among the issues raised, the most important seems to be that the data set was never validated with respect to mimicking real network traffic data. Examination has shown that data rates were far below what will be experienced in a real medium sized network. However, it was found that the data set is still useful to evaluate the true positive performance of an IDS but the researcher's conclusion pertaining to false-positive results were meaningless.

In case one compiles one's own data set for IDS modeling, using sensor data collected across a computer network, it is advisable to pay attention to the decentralised nature of the intrusion detection problem. That is, (i) relevant information may be present at multiple sources, (ii) relevant information may be present at multiple levels of abstraction, (iii) relevant information needs to be presented at the human's (system operator, administrator) level of understanding. It is necessary to be aware of the possibly heterogeneous nature of networks, where information is represented as electrical signal on the wire, as TCP/IP packet at the network interface, or some HTTP packet inside a browser, which is suitable for processing by computers but may be difficult to interpret by humans. Also, since practitioners of unauthorised intrusion are humans, any human might understand their habits, motivations, and actions. For this, Corona et. al present a conceptual framework to information fusion and how it fits into the current perception of computer security by individuals and cooperations [1]. The authors raise issues like: What data should be collected, how should it be collected and where (source, resolution, category)? How should data acquired from different sources be correlated (spatial, temporal, functional)? How can data be suitably abstracted into objects? How can objects be used to describe situations or state of a distributed environment using characteristics like context, dependency, origin, source or protocol? How can the situational knowledge about an environment and domain-expertise (rules) be used to assess the risk of an environment being intruded ? Should this data and information fusion be conducted centrally or in a distributed manner. Who and how can authority, reliability, confidentially and consent be ensured in such a scenario ?

Regardless of whether one uses a gold standard or a self-compiled data set, identifying whether or not (i) the data set accurately reflects the problem under observation and (ii) all features presented do influence the IDS result, might improve model building and performance. Mukkamala and Sung propose a rule-based feature subset selection approach implemented using the DK99C data set [25]. As noted, the motivation behind feature subset selection is that it does not sacrifice accuracy for reduced training time, there is a reduced risk of divergence during training, and an improved overall performance. The authors propose if-then rules evaluating measures like accuracy, training time and testing time to assess whether or not a feature is important, secondary or insignificant. Measures are collected from iteratively training a support vector machine (SVM) and ANN model on the DK99C data set. This means, for each feature and each of the four attack categories in the DK99C data set, the following steps were taken: feature was removed from training and test data, the respective classifier (ANN, SVM) was trained, the respective classifier was evaluated against a baseline classifier (same type but no removed features) using the mentioned measures, and collected measures and rules are used to rank the respective feature. After feature subsets had been determined, the authors ran a series of experiments comparing SVM and ANN classification performance with regard to using all features, important features per category or union of important features per category. The authors note that SVM performance was better than ANN performance with respect to accuracy (only very small difference), training time, testing time and scalability (number of training samples processed) regardless of selected features.

5.2 *Robustness and Performance*

As reported by several authors, BP learning when used with MSE training performance measure, may fail to converge if training data contains a high number of outliers [26]. For this, El-Melegy et. al propose different learning performance measures, such as maximum likelihood estimate (M-estimate) or least-median-squares estimate (LMedS-estimate). According to the authors, a first remedy may be to incorporate M-estimate into BP learning. Unfortunately, M-estimate delivers robust classification only if training data contains less than 30% outliers, otherwise classification results become unstable. For this, El-Melegy et. al propose using the LMedS estimate in ANN learning. Unfortunately, there is no known closed mathematical form describing LMedS estimate. Consequently, it cannot be integrated into the BP learning scheme. To resolve this situation, a substitute to BP learning, stochastic simulated annealing, is developed and proposed to be integrated with ANN learning. Stochastic simulated annealing, randomly tries to minimise the LMedS estimate error over the whole training data. The authors evaluated both M-estimate and LMedS-estimate against RMSE measure using the task of 1-D and 2-D function approximation. Results show that LMedS-estimate is superior to M-estimate and RMSE in both evaluation scenarios and LMedS-estimate is able to handle high amounts of outlier (up to almost 50%). Unfortunately, computing LMedS-estimate is computationally intense because of the randomly performed search for the global minimum LMedS-estimate. However, parallel computing might reduce LMedS-estimate computation time. Another improvement could be hybridisation of stochastic simulated annealing with deterministic algorithms to replace random behaviour with computationally less intense deterministic behaviour whenever possible. As noted, predictable low training and run-time of a classification system are important to transition from purely academic to real-world approaches and achieving acceptance among users. For this, El-Bakry and Hamada propose a novel implementation of ANN in Fourier space, to significantly reduce run-time performance of pattern recognition in images [27]. By performing cross-correlation operation in Fourier space instead of spatial space, the computationally expensive cross-correlation operation is transformed into a less costly matrix multiplication. Further performance improvement can be reached by assuming the input image has quadratic dimensions. A divide-and-conquer scheme as used by Fast Fourier Transformation, can be employed to segment the original (large) quadratic image into quadratic subregions of smaller extent. This subdivision enables exploitation of parallel processing thereby reducing overall run-time performance. Unfortunately, El-Bakry and Hamada's approach is limited to the domain of pattern recognition in images. Whether or not their approach can be extended to intrusion detection, has to be evaluated. Also, the author's main objective is reducing run-time, i.e. time needed to classify a certain input pattern, which although time-critical, is by an order of magnitude shorter than training time of the respective classifier. However, run-time performance is important with respect to real-time detection because model re-training / update could be performed on-demand asynchronously.

6 Conclusion

This chapter has presented an introduction to using NN to perform intrusion detection. We have looked at very different types of neural networks and looked at methods to identify specific intrusion types. It has been argued that NN are a viable approach to building robust and accurate IDS being able to work in noisy, incomplete and dynamic environments. It has been pointed out that key to a successful NN intrusion detection model is the choice of an appropriate data set reflecting implicit domain knowledge (size, dimensionality, data format, reflectance of phenomenon under study) and model-specific characteristics.

Although, intrusion detection research has advanced considerably, it is still far from having solved the problem of perfectly detecting unauthorised intrusions. Problems like integrating information from disparate sites in a heterogeneous environment, visualisation of intrusion detection results, recommending appropriate actions and counter-measures to either prevent or resolve intrusions, providing real-world deployable IDS obeying legal regulations/corporate security requirements/service-level-agreements, or reaching acceptance among professionals still persist.

No matter what the choice of intrusion detection model is going to be, computer security in general and intrusion detection in particular is still an open, diverse and challenging field with enormous potential due to the ever-expanding nature of communication networks.

Reference

1. Corona, I., Giacinto, G., Mazzariello, C., Roli, F., Sansone, C.: Information fusion for computer security: State of the art and open issues. Information Fusion 10(4), 274–284 (2009); Special Issue on Information Fusion in Computer Security
2. Alfantookh, A.A.: Dos attacks intelligent detection using neural networks. Journal of King Saud University. Computer & Information Sciences 18(1), 27–44 (2005)
3. Barika, F., Hadjar, K., El-Kadhi, N.: Artificial neural network for mobile ids solution. In: Security and Management, pp. 271–277 (2009)
4. Young, G., Pescatore, J.: Magic quadrant for network intrusion prevention system appliances. Gartner Core RAS Research Note G00167303, pp. 1–12 (April 2009)
5. Bishop, C.: Neural Networks for Pattern Recognition. Clarendon Press (1995)
6. Duda, R.O., Hart, P.E., Stork, D.G.: Pattern Classification, 2nd edn. Wiley Interscience (2000)
7. Jiawei, H., Kamber, M.: Data Mining: Concepts and Techniques, 2nd edn. Morgan Kaufmann Publishers Inc., San Francisco (2006)
8. Iftikhar, A., Abdullah, A.B., Alghamdi, A.S.: Application of artificial neural network in detection of dos attacks. In: SIN 2009: Proceedings of the 2nd International Conference on Security of Information and Networks, pp. 229–234. ACM, New York (2009)
9. Beyer, W., Liebscher, M., Beer, M., Graf, W.: Neural network based response surface methods - a comparative study. 5th German LS-DYNA Forum, pp. 29–38 (October 2006)
10. Zhu, B., Shajari, M., Onut, I.V., Kabiri, P.: Evaluation Criteria. In: Network Intrusion Detection and Prevention Concepts and Techniques. Advances in Information Security, vol. 47, pp. 161–182. Springer, Heidelberg (2010)
11. Chang, R.I., Lai, L.B., Su, W.D., Wang, J.C., Kouh, J.S.: Intrusion detection by backpropagation neural networks with sample-query and attribute-query. International Journal of Computational Intelligence Research 3(1), 6–10 (2007)
12. Wu, C.H.: Behavior-based spam detection using a hybrid method of rule-based techniques and neural networks. Expert Systems With Applications 36(3, Part 1), 4321–4330 (2009)
13. Herrero, A., Corchado, E., Pellicer, M.A., Abraham, A.: Hybrid multi agent-neural network intrusion detection with mobile visualization. Innovations in Hybrid Intelligent Systems 44, 320–328 (2007)

14. Alonso-Betanzos, A., Sánchez-Maroño, N., Carballal-Fortes, F.M., Suárez-Romero, J.A., Pérez-Sánchez, B.: Classification of computer intrusions using functional networks - a comparative study. In: ESANN, pp. 579–584 (2007)
15. Stopel, D., Boger, Z., Moskovitch, R., Shahar, Y., Elovici, Y.: Application of artificial neural networks techniques to computer worm detection. In: International Joint Conference on Neural Networks (IJCNN), pp. 2362–2369 (2006)
16. Bai, J., Wu, Y., Wang, G., Yang, S.X., Qiu, W.: A Novel Intrusion Detection Model Based on Multi-Layer Self-Organizing Maps and Principal Component Analysis. In: Wang, J., Yi, Z., Żurada, J.M., Lu, B.-L., Yin, H. (eds.) ISNN 2006. LNCS, vol. 3973, pp. 255–260. Springer, Heidelberg (2006)
17. Mitrokotsa, A., Douligeris, C.: Detecting denial of service attacks using emergent self-organizing maps. In: International Symposium on Signal Processing and Information Technology, pp. 375–380 (2005)
18. Calado, J., da Costa, J.S.: Fuzzy Neural Networks Applied to Fault Diagnosis. Advanced Information and Knowledge Processing. In: Computational Intelligence in Fault Diagnosis, pp. 305–334. Springer, London (2006)
19. Shafi, K., Abbass, H.: Biologically-inspired complex adaptive systems approaches to network intrusion detection. Information Security Technical Report 12(4), 209–217 (2007)
20. Karim, A.: Computational Intelligence for Network Intrusion Detection: Recent Contributions. In: Hao, Y., Liu, J., Wang, Y.-P., Cheung, Y.-m., Yin, H., Jiao, L., Ma, J., Jiao, Y.-C. (eds.) CIS 2005. LNCS (LNAI), vol. 3801, pp. 170–175. Springer, Heidelberg (2005)
21. Herrero, Á., Corchado, E.: Traffic Data Preparation for a Hybrid Network IDS. In: Corchado, E., Abraham, A., Pedrycz, W. (eds.) HAIS 2008. LNCS (LNAI), vol. 5271, pp. 247–256. Springer, Heidelberg (2008)
22. Mukkamala, S., Sung, A.H., Abraham, A.: Intrusion detection using an ensemble of intelligent paradigms. Journal of Network and Computer Applications 28(2), 167–182 (2005); Computational Intelligence on the Internet
23. Yin, H.: The Self-Organizing Maps: Background, Theories, Extensions and Applications. In: Computational Intelligence: A Compendium. SCI, vol. 115, pp. 715–762. Springer, Heidelberg (2008)
24. Brugger, T.: Kdd cup '99 dataset (network intrusion) considered harmful (September 2007), http://www.kdnuggets.com/news/2007/n18/4i.html (last accessed November 19, 2009)
25. Mukkamala, S., Sung, A.H.: Identifying significant features for network forensic analysis using artificial intelligence techniques. International Journal of Digital Evidence (IJDE) 1(4), 1–17 (2003)
26. El-Melegy, M.T., Essai, M.H., Ali, A.A.: Robust Training of Artificial Feedforward Neural Networks. In: Hassanien, A.-E., Abraham, A., Vasilakos, A.V., Pedrycz, W. (eds.) Foundations of Computational, Intelligence Volume 1. SCI, vol. 201, pp. 217–242. Springer, Heidelberg (2009)
27. El-Bakry, H.M., Hamada, M.: A New Implementation for Neural Networks in Fourier-Space. In: Hassanien, A.-E., Abraham, A., Vasilakos, A.V., Pedrycz, W. (eds.) Foundations of Computational, Intelligence Volume 1. SCI, vol. 201, pp. 307–330. Springer, Heidelberg (2009)

Evolutionary Computation in Computer Security and Forensics: An Overview

Sulaiman Al amro, David A. Elizondo, Agusti Solanas,
and Antoni Martínez-Ballesté

Abstract. In daily life, people are faced with problems that have more than one solution. In computing, and other fields, researchers and developers encounter computational problems that may have a large number of solutions. In these cases, it may be almost unfeasible to determine the best one. Evolutionary computation is an appropriate technique for finding an optimal solution to a problem. Among these problems, evolutionary computation can provide an efficient way to solve those associated with computer security and forensics. This chapter examines the approaches evolutionary computation offers to discover an optimal solution to a problem. Moreover, it overviews how evolutionary computation can be applied to different scenarios related to computer security and forensics.

1 Introduction

From the many techniques that have been proposed to investigate problems, algorithms that rely on the principles of natural selection are among those that

Sulaiman Al amro
Centre of Software Technology Research Laboratory (STRL), De Montfort University,
The Gateway, Leicester, LE1 9BH, United Kingdom
e-mail: salamro@dmu.ac.uk

David A. Elizondo
Computer Forensics and Security Group, Department of Computer Technology,
De Montfort
University, The Gateway, Leicester, LE1 9BH, United Kingdom
e-mail: elizondo@dmu.ac.uk.

Agusti Solanas · Antoni Martínez-Ballesté
Dept. Computer Engineering and Maths, Universitat Rovira i Virgili,
43001 Tarragona, Catalonia, Spain
e-mail: {agusti.solanas,antoni.martinez}@urv.cat

D.A. Elizondo et al. (Eds.): Computational Intelligence for Privacy and Security, SCI 394, pp. 25–34.
springerlink.com

have resulted in optimal solutions [1]. The hypothesis of natural selection suggests that today's animals and plants are the outcome of millions of years of modifications to adapt to the requirements of the environment. Organisms compete for environmental resources, and those that are best adapted to exploit these resources will reproduce more successfully; therefore their offspring will be more numerous [2]. It is well known that each new organism is better adapted than its predecessors because it inherits characteristics that tend to promote the continuation of life. Evolutionary computation (EC) is inspired by this process of natural evolution and expresses its rules in algorithms that can be used to search through potential solutions to find which is best suited to resolve a particular problem. Thus, evolutionary computation imitates natural evolution, but on a computer, via a series or sequence of optimisation algorithms that usually depend on a set of rules. The optimisation process iteratively enhances the quality of solutions until a better one is achieved [1]. Therefore, evolutionary computation does not derive its ideas from human intelligence, but rather from the way organisms adapt from one generation to the next.

2 Background on Evolutionary Computation

Evolutionary computation is an important area on which many writers and researchers have produced numerous articles, books and studies due to the fact that it has many benefits and promises better solutions [8]. Evolutionary computation is becoming one of the most rapidly growing areas of computer science and engineering because of its ability to solve problems that previous algorithms could not handle [2]. There are two main approaches to EC, theoretical and empirical [2]. The theoretical approach is used to search through algorithms to seek their correspondence to mathematical truths, while the empirical approach is used to examine evolutionary algorithms by statistical means. Alternatively, by creating a population of individuals, appraising their fitness, producing a new population via genetic procedures and repeating this process several times, all EC methods imitate the natural evolutionary processes of selection, mutation and reproduction [1]. Ultimately, there are three main methods of evolutionary computation [3]: genetic algorithms, evolutionary strategies and genetic programming.

The evolutionary computation approach has several advantages over traditional methods[2], for example:

- The ability to adapt to changing situations. In most traditional approaches, if a problem occurs, the process must be restarted from the beginning, whereas EC approaches can be adjusted to restart from a particular point when a problem arises.
- A second advantage is its ability to provide solutions as quickly as possible.

During the rest of this section we review different approaches related to EC: genetic algorithms, evolution strategies and genetic programming.

2.2 Genetic Algorithms

Genetic algorithms (GAs) are random search algorithms which mimic the ability of biological evolution to provide better solutions to given problems by moving from one population of artificial chromosomes to another [1]. To replace one population with another, reproduction is achieved by measuring each chromosome's fitness. This process of reproduction can be achieved by two genetic operators, crossover and mutation. Crossover is used to trade off parts of two single chromosomes. Inside the chromosome there are a number of gene values and mutation is used to change some of these in order to achieve the best solution [1]. However, given a problem to be solved and a string of binary representations for candidate solutions, a genetic algorithm can be achieved by following the steps illustrated in Figure 1 (left), beginning with a population of randomly generated chromosomes [3]. The fitness of each chromosome in the population is then calculated. Next, a couple of parent chromosomes are selected from the current population, depending on their fitness. The fourth step is to implement the genetic operators (crossover and mutation) to create a couple of offspring chromosomes, which are then inserted into a new population. This step is repeated until the size of the new population is equal to the first, when the old one is replaced by the new. Finally, the process is repeated through the calculation of each chromosome's fitness, up to the point when a satisfactory result is achieved. In genetic algorithms, selection is a genetic operator that is able to choose a chromosome in the present generation's population and include it in the following generation's population.

2.3 Evolution Strategies

Evolution strategy (ES) is another approach to EC that simulates natural evolution and differs from GA by using just one operator: a mutation [1]. This operator is used to generate one offspring from one parent per generation, as shown in Figure 1 (right). This can be achieved through the following set of steps. First, the problem is presented by choosing a number of parameters and subsequently determining the appropriate range of each parameter. Secondly, for each parameter in the range, an initial value is selected.

These values are used as the initial population of parent parameters. Thirdly, a solution is calculated from the parent parameters. The fourth step is to create new parameters, which are called offspring parameters, by means of a mutation operator. Mutation operator, known as the main search procedure in ES, randomly creates samples for solution candidate that is selected from previous and different search points. Solution is then calculated from the offspring parameters and compared with the solution from step three. If the new solution is better, then it replaces the old one; otherwise, the parent parameter solution is kept. Finally, step four is repeated until the solution is satisfactory, or until a specified number of generations are achieved.

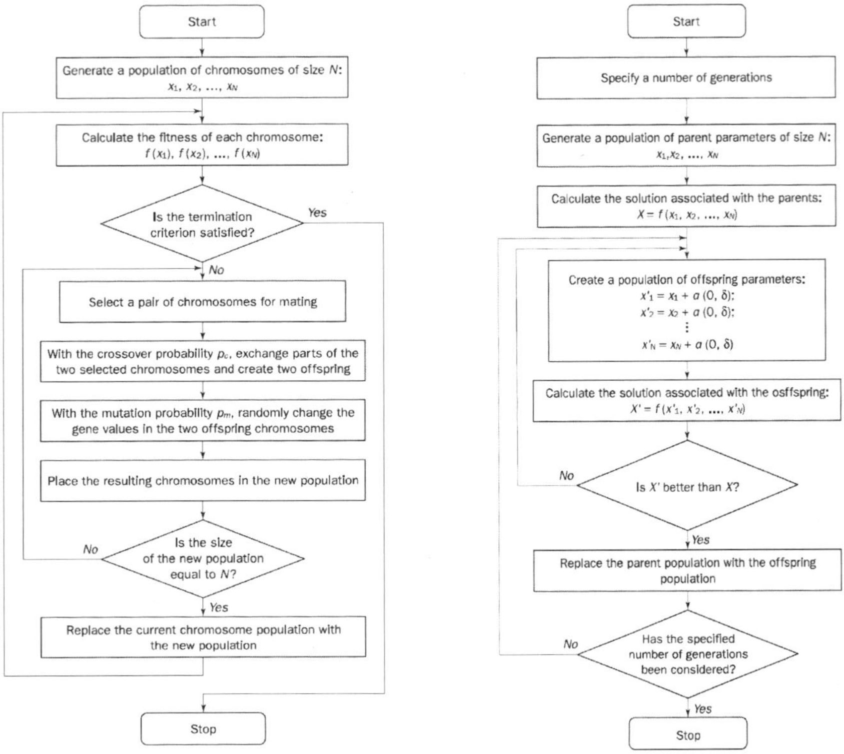

Fig. 1 Left, a basic genetic algorithm [1]. Right, block diagram of the (1+1) evolution strategy [1].

2.4 Genetic Programming

Genetic programming (GP) is "a domain-independent method relying on genetic breeding of populations of computer programs to automatically create a computer program from high-level requirements" [4]. In addition, GP can be seen as a development of GA which does not just produce a string of bits to solve the problem but also generates a computer code for problem-solving. That is, while GA are used to solve problems by producing a set of binary numbers, GP uses a computer program as the solution [1]. Any program in a computer has a sequence of functions referred to values, but because different programming languages have their own statements and operations and also have disparate syntactic restrictions, the role of GP is to manipulate programs by implementing genetic operators. This allows the program to be manipulated as a data and the new data to be executed as a program. However, before GP is applied to a problem, five steps must be accomplished [4]. The first step is to determine the set of terminals. Secondly, select the set of primitive functions. Next, define the fitness function and decide on the parameters for controlling the run. Finally, choose the method for designating a result of the run. Different methods of evolutionary computation can produce solutions for given problems which take different shapes, depending on the problem and the method used to solve it.

3 Evolutionary Computation and Related Approaches to Security

Evolutionary computation techniques (such as GA and GP) are of increasing interest in the computer security area, in cryptology and in both network and host security, because traditional techniques for computer security have not provided a full set of solutions in these fields [5]. This is the case of the design and implementation of Intrusion Detection Systems (IDS). Furthermore, many algorithms have been applied to many fields in the area of computer security and also have proven efficacy in detecting malicious attacks on network systems, analysing and designing several cryptographic primitives, making block ciphers, etc. [5]. Hence, evolutionary computation offers solutions to security problems which could not be provided by previous techniques.

3.1 Selected Applications of Genetic Algorithms to Computer Security

Genetic algorithms provide an efficient way to solve problems associated with computer security, especially in detecting intrusions and malicious attacks [6]. Network traffic can be subjected to a set of rules that determine whether the network connections and related behaviours are intrusions or not. These rules are modelled by the chromosomes in the population, and until the evolution criteria are established, the population will evolve [7]. Inside an IDS, a set of rules are used to decide if the network connection and associated behaviours are encroachments or not; these rules can be presented as [6]:

If {*condition*} then {*act*}

When a connection appears, the *condition* compares this with the rules in the IDS to determine the possibility of intrusion. If an intrusion has occurred, the *act* will engender a reaction which has been identified by the security department in the organisation, such as alerting the security administrator or stopping the connection. The evaluation function, which is one of the most important parameters in GA, is used to computing outcome and fitness by calculating the goodness of each chromosome. In addition, a number of parameters should be considered during the implementation, such as the crossover rate, the mutation rate, the population numbers and the generation numbers. These parameters should be set in accordance with the environment of the system and the security policies of the organisation [7].

GA-based IDS can be achieved by having two modules, each of which has two stages [9]:

- The first is the training stage, where a set of classification rules are created using a genetic algorithm from network audit data.
- The second is the intrusion detection stage, where the rules that have been generated are used to group arriving network connections in the environment. Once the rules have been generated, the task of the IDS is simple and efficient. It starts to compare the incoming data with the rules and determines whether the connections and related behaviours are intrusions or not [9].

Other GA techniques can also be used with an IDS to overcome some of the problems that arise when traditional methods are used. With if-then rules in expert systems, an intrusion might not be detected if it deviates slightly from a pattern extracted from data, or a false alarm might be caused if a minor change in normal behaviour occurs [10]. However, this problem can be solved by integrating a genetic algorithm with clustering analysis for the purpose of intrusion detection. In this case, an IDS-based clustering genetic algorithm is composed of two stages: clustering analysis and genetic optimising. In the first stage, initialised clustering sets are constructed by resemblance rules, then genetic optimising allows them to identify normal and abnormal actions in the system [10]. The genetic optimising algorithm is a randomisation search technique that is used in optimisation problems. Its main aim in this approach is to output the optimizing clustering set.

3.2 Selected Application of Genetic Programming to Computer Security

In order to train and generate rules that help to detect malicious attacks on a network, a GP-based intrusion detection network uses audit data and audit trails comprising features which are associated with connection data [11]. However, GP has the ability to learn new events in the network and to use computer programs to detect malicious attacks. These properties make it more powerful in the area of computer security and intrusion detection [12].

An IDS requires the ability to modify itself to suit new information (e.g. new attack patterns). Therefore, since a system that is able to evolve new agents and train novel ones is needed, genetic programming can be used. It is important that the genetic programs are written in a simple language that allows it to gain entrance to the audit data field and process them. An evaluator interprets these genetic programs and provides them with audit information to give a better solution [12].

Linear genetic programming (LGP) is a GP variant that evolves sequences of instructions from an imperative programming language [13, p. 6]. LGP has been successfully applied to diverse machine learning problems and can be used in IDS to distinguish the key features that assist in making a decision as to whether a connection is subject to normal or abnormal activity [14]. It uses computer programmes that track the connection and related behaviour by monitoring the key features for intrusion detection purposes, determining whether this behaviour threatens the network system [14]. Hence, GP can be implemented for the purpose of intrusion detection by using computer programmes that monitor the network traffic and detect malicious attacks. It can also recognise unknown attacks, because it has the ability to adapt to changes in the environment to detect new attack patterns.

4 Computer Forensics Applications Using Evolutionary Computation

Evolutionary computation algorithms have been used efficiently within a forensics system to gather network forensics data and to track novel attacks. In addition, they can be used to search and gather network forensics from a large amount of network data and have provided valuable reports on malicious behaviour in such networks. Indeed, the robustness and scalability of EC make it an appropriate technique to analyse and track evidence in the network [15]. The idea behind gathering network forensics using EC is to create a formal characterization of the information required and generated in each stage in the investigation and to produce a network forensics tool in which the standard tools in each of the stages of an investigation are integrates.

Automated face recognition and eye detection have become important applications in many fields and can be used in criminal investigations. Evolutionary computation methods such as genetic algorithms can be used successfully within systems designed to provide automated face recognition and eye detection. This can be achieved by using a hybrid approach, integrating the learning and evolutionary abilities of EC, in order that the systems should be able to evolve and learn while in use [16]. In addition, EC has the ability to search for optimal solutions, which is one of the most interesting abilities that face recognition need. Therefore, EC can be used in automated face recognition and eye detection to help in the investigation of malicious attacks and its learning capacity makes such a system robust over time.

Forensic tools are very useful in data collection and analysis, but that they are not appropriate in investigating insider attacks, because these are very complex to deal with [8]. Therefore, very good guidance plans are required in order to achieve attack attribution [17]. One of the most useful approaches here is to use the capability of genetic algorithms to create a set of optimised rules that recognise unauthorised processes and perform function-based process verification [18]. Moreover, the strong classification ability of GA is useful not only for intrusion detection, but also – and more significantly – for network forensics. This ability helps forensic investigators because it provides data analysis for the exact type of threat and the way it is presented [19].

One of the most important aspects of software forensics [20] is author identification, which determines the probability of a novel fragment of code having been written by a particular programmer, through comparing specimens of code produced by a large number of programmers [21]. This can be achieved by using the three basic parts of GA, namely, genes, organisms and populations, as detailed next. First, genes provide the individual problems that will be optimised. A set of metrics active in the classifier is optimised in order to increase the precision of the classification. The classifier is interfaced to the genetic algorithm so that, instead of outputting a ranked list of authors for each project in both testing sets, the classifier produced a single list identifying the number of correct classifications at each level of precision. The candidate solution of the problem is then represented by an organism subclass. Each subclass will have a fitness function which appraises the quality of the solution compared to others. Each execution of a fitness function corresponds to a full test run of the classifier. Finally, a set of organisms is randomly selected to comprise the population. The population class contains and directs the organisms to reproduce. The operational parameter of the population allows the operation of the GA to be adjusted. A number of operational parameters are generated in order to control and modify the genetic algorithm behaviour without requiring significant additional programming.

Another problem arisen in computer forensics is how to determine the type of a file fragment. This is not needed when it contains the initial header, which precisely identifies the file type, but is essential when the fragment is from the middle of the file [22]. This problem can be solved by using a genetic algorithm, because it can provide a better classification solution than traditional techniques, as follows. First, a random population of prediction functions is created, then these compete and reproduce, based on their accuracy in predicting the type of a given file fragment. Over time, each population will have better predictors than the previous one. Finally, the best of them are used to replace the poorest ones in order to obtain the best predictors. In this way, GA has shown some degree of success in determining file types from large quantities of data. Furthermore, there is a problem which may occur occasionally, where two types of file have many similar data characteristics: It is extremely difficult to determine the file type by traditional techniques, but the GA approach has been shown to deal very well with this problem [22].

Genetic algorithms can also be used in other areas of forensics and can save investigators time. A process called "craniofacial superimposition", described in [23], helps to identify an individual by coating a picture on a skull model. This process is usually effected manually by forensic scientists and is therefore very difficult and time consuming, but GA can provide an automated mechanism that helps investigators to determine automatically and rapidly which skull belongs to whom. First, a set of population solutions is generated to try to determine who the person is. Next, successive generations are constructed by using genetic operators (mutation, crossover and reproduction). Finally, the forensic expert analyses the similarities between the picture and the skull and makes decisions (positive, negative or unknown) by examining the best superimposition. Using GA has provided a fast and accurate method, and thus very applicable for solving one of the most tedious works performed by the forensic research [23].

5 Conclusions

In this chapter we have overviewed the use of evolutionary computation in some security and forensics applications. The concept of evolutionary computation is derived from the evolution that takes place in nature, which takes the advantages gained from one generation and passes them on to the next. Evolutionary computation aims at choosing the best or optimal solution to a problem. It has proved to be a good choice for computer security and has yielded various algorithms that deal with different scenarios, such as IDS or file type recognition.

References

[1] Negnevitsky, M.: Artificial Intelligence: A guide to intelligent systems. Pearson Education Limited, Essex (2005)

[2] Fogel, D.: Evolutionary Computation Toward a New Philosophy of Machine Intelligence, 3rd edn. IEEE Press, Piscataway (2006)

[3] Mitchell, M.: An Introduction to Genetic Algorithms. MIT Press, Cambridge (1998)

[4] Koza, J., Bennett, F., Andre, D., Keane, M.: Genetic programming III: Darwinian invention and problem solving. Kaufmann Publishers (1999)

[5] Castro, J., Vinela, P.: Expert Evolutionary Computation in Computer Security and Cryptography. New Generation Computing Magazine 23(3), 193–199 (2005)

[6] Sinclair, C., Pierce, L., Matzner, S.: An Application of Machine Learning to Network Intrusion Detection. In: Computer Phoenix, AZ, pp. 371–377. IEEE, AZ (1999)

[7] Li, W.: Using Genetic Algorithm or Network Intrusion Detection. In: Proceedings of the United States Department of Energy Cyber Security Group 2004 Training Conference (CD ROM) CiteSeerX, p. 8. US Department of Energy CSG, Kansas city (2004)

[8] Al- Amro, S.: Computational Intelligence towards Computer Security and Forensics. MSc thesis, Leicester,De Montfort University (2009)

[9] Gong, R., Zulkernine, M., Abolmaesumi, P.: 'A Software Implementation of a Genetic Algorithm Based Approach to Network Intrusion Detection. In: Proceedings of the Sixth International Conference on Software Engineering, Artificial Intelligence, Networking and Parallel/Distributed Computing and First ACIS International Workshop on Self-Assembling Wireless Networks, pp. 246–253. IEEE, Washington, DC (2005)

[10] Zhao, J.L., Zhao, J.F., Li, J.: Intrusion Detection based on Clustering Genetic Algorithms. Machine Learning and Cybernetics 6, 3911–3914 (2005)

[11] LaRoche, P., Zincir-Heywood, A.: 802.11Network Intrusion Detection using Genetic Programming. In: Proceedings of the 2005 Workshops on Genetic and Evolutionary Computation, pp. 170–171. ACM, New York (2005)

[12] Crosbie, M., Spafford, G.: Applying Genetic Programming to Intrusion Detection. In: Proceedings of the AAAI 1995 Fall Symposium Series on Genetic Programming, pp. 1–8. AAAI, Menlo Park (1995)

[13] Brameier, M., Banzhaf, W.: Linear Genetic Programming. Springer Science + Business Media, LLC, New York (2006)

[14] Mukkamala, S., Sung, A., Abraham, A.: 'A Modeling Intrusion Detection Systems Using Linear Genetic Programming Approach'. In: Proceedings of the 17th International Conference on Innovations in Applied Artificial Intelligence, pp. 633–642. Springer, Heidelberg (2004)

[15] Merkle, L.: Automated Network Forensics. In: Proceedings of the 2008 GECCO Conference Companion on Genetic and Evolutionary Computation, pp. 1929–1932. ACM, New York (2008)

[16] Huang, J., Liu, C., Wechsler, H.: Eye Detection and Face Recognition Using Evolutionary Computation. In: Proceedings of NATO-ASI on Face Recognition: From Theory to Applications, pp. 348–377. Springer, Heidelberg (1998)

[17] Upadhyaya, D., Ngo, H., Pramanik, S., Chinchani, R., Mathew, S.: Insider Threat Analysis Using Information-Centric Modeling. In: Craiger, P., Shenoi, S. (eds.) Advances in Digital Forensics, 3rd edn., pp. 53–73. Springer, New York (2007)

[18] Bradford, P., Hu, N.: A layered approach to insider threat detection and proactive forensics. In: Technology Blitz of the Twenty-First Annual Computer Security Applications Conference, Tucson, AZ. ACSAC, Silver Spring (2005)

[19] Bankovic, Z., Stepanovic, D., Bojanic, S., Nieto-Taladriz, O.: Improving network security using genetic algorithm approach. Journal of Computers and Electrical Engineering 33(1), 438–451 (2007)

[20] Lange, R., Mancoridis, S.: Using Code Metric Histograms and Genetic Algorithms to Perform Author Identification for Software Forensics. In: Proceedings of the 9th Annual Conference on Genetic and Evolutionary Computation, pp. 2082–2089. ACM, New York (2007)

[21] Gray, A., Sallis, P., MacDonell, S.: A Dictionary-based System for Extracting Source Code Metrics for Software Forensics. In: Proceedings of Software Engineering: Education & Practice, pp. 252–259. IEEE Computer Society, California (1998)

[22] Calhoun, W., Coles, D.: Predicting the types of file fragments. The Journal of Digital Investigation 5(1), 14–20 (2007)

[23] Ballerini, L., Cordon, O., Santamaria, J., Damas, S., Aleman, I., Botella, M.: Craniofacial Superimposition in Forensic Identification using Genetic Algorithms. In: The Third International Symposium on Information Assurance and Security, pp. 429–434. IEEE Computer Society, California (2007)

Application of Fuzzy Logic in Computer Security and Forensics

S. Al Amro, F. Chiclana, and D.A. Elizondo

Abstract. Due to the growing reliance that corporations and government agencies place on their computer networks, the significance of defending these systems from attack cannot be underestimated. A single malicious encroachment into a computer network can cause a great deal of damage to an organisation. Computer Security concerns the protection of information and property from misuse by unauthorized people. Computer forensics provides methods for the preservation, collection, validation, identification, analysis, interpretation, documentation and presentation of digital evidence derived from digital sources for the reconstruction of criminal activities or policy violation. Fuzzy logic can properly be used to help deal with issues associated with computer security and computer forensics. This chapter deals with the use of fuzzy logic in computer security and forensics, providing a general overview of some of its applications, recent algorithms, abilities, advantages and drawbacks.

1 Introduction

In spite of the fact that human beings do not have sufficient capacity to deal accurately with large amounts of numerical information and data, they are extremely

S. Al Amro
Software Technology Research Laboratory (STRL), Department of Computer Technology, De Montfort University, Leicester, United Kingdom
e-mail: salamro@dmu.ac.uk

F. Chiclana
Centre for Computational Intelligence, Dept. of Informatics, De Montfort University, Leicester, United Kingdom
e-mail: chiclana@dmu.ac.uk

D.A. Elizondo
Computer Forensics and Security Group, Department of Computer Technology, De Montfort University, Leicester, United Kingdom
e-mail: elizondo@dmu.ac.uk

D.A. Elizondo et al. (Eds.): Computational Intelligence for Privacy and Security, SCI 394, pp. 35–49.
springerlink.com

skilful at making complex decisions. The opposite is true of computers, which can carry out highly complex calculations in a fraction of a second, while being totally unable to recognise the most basic activities of humans unless they are numerically represented. This evident mismatch between the abilities of human beings and of numerical systems led to the development the theory of fuzzy logic [1], which has its origins in Zadeh's seminal paper 'Fuzzy Sets' published in 1965 [2]. Recent developments of this logic have affected most aspects of modern technology, as Japanese researchers have increasingly incorporated it into their products and industries. Thus, it has become customary to see air conditioners, cameras, washing machines and other items of equipment which use some kind of control system based on fuzzy logic (see [3] for more details).

The theory of fuzzy logic has managed to fill some significant gaps in the existing classical logic, because the latter relies on quantitative approaches to system analysis or decision making. However, if accuracy is both required and feasible when dealing with simple systems or decisions, it is often not possible and sometimes not required when dealing with complex issues. In addition, the more complex a system becomes, the less usefulness there is in exact statements and the less accuracy there is in useful statements. The difference between fuzzy and crisp logic is a philosophical one which depends on the dialectic of significance and precision [3]. Thus, not everything accurate (obvious) is important, and not every important fact can be stated precisely. For example, if you observe a dangerous animal, object or event about to harm a friend, it is not important to be precise in describing the danger to him/her, but only to warn him/her to avoid this threat. This is an example of many real cases where precision is deadly and vagueness is essential to survival; it illustrates what fuzzy logic is about.

2 Fuzzy Logic

Classical sets only allow full membership or no membership at all [4]. An example of such set is the 'class of all real numbers which are greater than 1'. It is obvious that in this case there is a precise criteria of membership, i.e. given any number there is no ambiguity status with respect to this class and therefore we have a so-called 'black-or-white' membership. In most cases, however, the precision required in the definition of a classical set is inappropriate. Take for example the 'class of all real numbers which are much greater than 1'. It is clear that ambiguity arises in the case of a number such as 10 in relation to this class. We could assert that this is no class or set in the usual mathematical sense of these terms, unless a threshold value is introduced to force a black-or-white membership. This type of *imprecisely* defined classes play an important role in human thinking [2].

In classical set theory there is a gap between the mathematical and mental representation of reality. Mathematical representations are based on *binary logic, precise numbers* ..., and as a consequence we have that classical logic is too rigid to account for gradual membership. On the other hand, mental representation is based on natural language terms such as *high temperature, young man, big size* ..., which are pervaded with vagueness in that there is a lack of specificity of linguistic terms and/or a lack of well defined boundaries of the class of objects they refer to.

Zadeh proposed the notion of a "Fuzzy Set" to fill this gap [2]. The word *fuzzy* stands for the word *vague*. Vagueness concerns the meaning of signs of a language. A concept is said to be vague when its meaning is not fixed by sharp boundaries. For so-called measurable concepts, we refer to (numerical) imprecision, although uncertainty is also used in this context. Fuzzy sets allow partial membership. In other words, an element may partially belong to a set. Therefore, results are not restricted to the absolutely true or false, but can be true or false to some degree. Fuzzy Set Theory aims to provide a systematic framework for dealing with the vagueness and imprecision inherent in human thought processes. Fuzzy logic is able to use vagueness to achieve a better answer, one that is more suited to the situation [5].

2.1 Fuzzy Sets

A classical set can be described in three different ways:

- Enumerating the elements that belong to the set (finite sets)
- Describing the set analytically by stating conditions for membership

$$A = \{x|property(x)\}$$

- Using the characteristic function, in which 1 indicates membership and 0 non-membership

$$\delta_A \colon X \rightarrow \{0, 1\}$$

$$\delta_A(x) = \begin{cases} 1 \text{ if } x \in A; \\ 0 \text{ if } x \notin A. \end{cases}$$

where X a universal set defined in a specific problem

For a fuzzy set, the characteristic function allows various degrees of membership for elements of a given set.

Definition 1 (Fuzzy Set). If X (universal set defined in a specific problem) is a collection of objects denoted generically by x, then a fuzzy set A in X is a set of ordered pairs:

$$A = \{(x, \mu_A(x))|x \in X\}$$

where $\mu_A \colon X \rightarrow [0, 1]$ is called the membership function of A and $\mu_A(x)$ represents the degree of membership of the element x in A.

Clearly, when the range of the membership function reduces to $\{0, 1\}$ we get the classical set.

Example 1 (Real numbers close to 10).

(a) Suppose $X = \{5, 6, 7, 8, 9, 10, 11, 12, 13, 14, 15\}$ then this fuzzy set could be modelled as:

$$\begin{aligned} A &= \{(5,0),(6,0.2),(7,0.4),(8,0.6),(9,0.8),(10,1),\\ &\qquad (11,0.8),(12,0.6),(13,0.4),(14,0.2),(15,0)\}\\ &= 0/5+0.2/6+0.4/7+0.6/8+0.8/9+1/10+\\ &\qquad 0.8/11+0.6/12+0.4/13+0.2/14+0/15 \end{aligned}$$

In general, in these cases a fuzzy set is represented as:

$$A = \sum_{i=1}^{n} \mu_A(x_i)/x_i$$

(b) Suppose $X = \mathbb{R}$ then this fuzzy set could be modelled as:

$$A = \{(x, \mu_A(x)|x \in X\}$$

$$\mu_A(x) = \frac{1}{1+(x-10)^2}$$

In these cases, the alternative notation is used:

$$A = \int_{\mathbb{R}} \frac{1}{1+(x-10)^2} \Big/ x$$

In general, in these cases the fuzzy set is denoted as:

$$A = \int_X \mu_A(x)/x$$

The representation of the second fuzzy set is given in Figure 1.

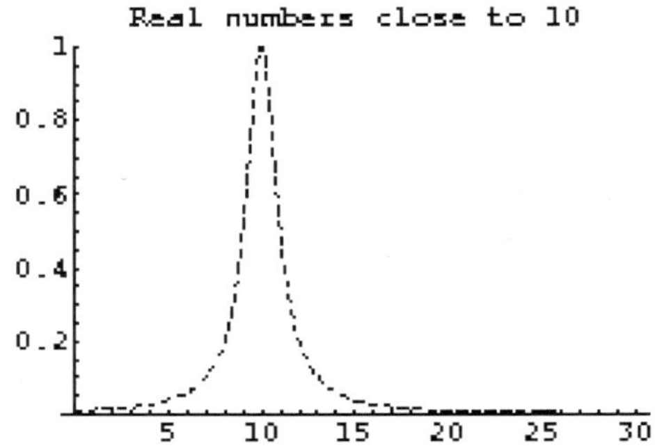

Fig. 1 Real numbers close to 10

Note that in the above notations, "/" does not refer to a division and is used as a symbol to separate the membership of an element from the element itself. A fuzzy set is completely characterised by its membership function, and therefore it is often understood as a function. The most usual membership functions used to represent fuzzy sets are the following:

- Symmetric triangular-shaped function: $A(x) = \begin{cases} b\left(1 - \frac{|x-a|}{s}\right), & a-s \leq x \leq a+s \\ 0, & otherwise \end{cases}$
- Trapezoidal shaped function: $A(x) = \begin{cases} \frac{(a-x)\cdot e}{a-b}, & a \leq x \leq b \\ e, & b \leq x \leq c \\ \frac{(d-x)\cdot e}{d-c}, & c \leq x \leq d \\ 0, & \text{otherwise.} \end{cases}$
- Bell shaped function: $A(x) = c \cdot e^{-\frac{(x-a)^2}{b}}$

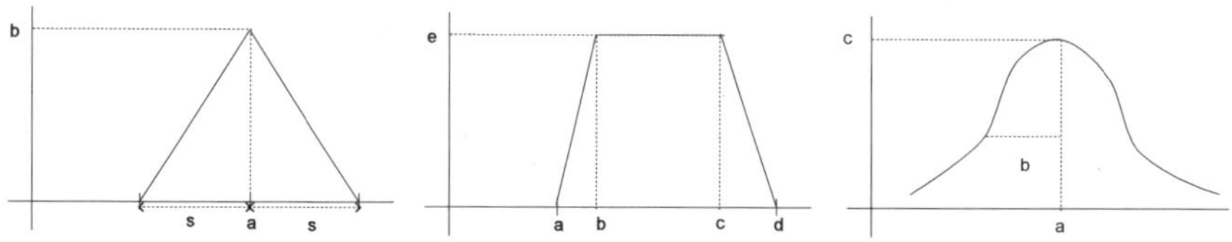

Fig. 2 Usual membership functions

2.2 *Linguistic Variables*

A linguistic variable is *"variable whose values are not numbers but words or sentences in a natural or artificial language"* [6, 7, 8]. The main purpose of using linguistic values (words or sentences) instead of numbers is that linguistic characterisations are, in general, less specific than numerical ones, but closer to the way humans express and use their knowledge. For example, "the building is tall" is less specific than "the building measures 300 meters". In this case, "tall" can be seen as a linguistic value of the variable "height" which is less precise and informative than the numerical value "300". Despite its less informative nature, the value "tall" allows humans to naturally express and deal with information that may be uncertain or incomplete (the speaker may not know the exact building height). Because situations where information is not precise are very common in real life, linguistic variables can be a powerful tool to model human knowledge.

Formally, a linguistic variable is a 5-tuple $\langle L, T(L), U, S, M \rangle$ [6] in which

- L is the name of the variable,
- $T(L)$ is a finite term set of labels or words (a collection of linguistic values),
- U is a universe of discourse,
- S is the syntactic rule which generates the terms in $T(L)$, and
- M is a semantic rule which associates with each linguistic value X its meaning $M(X)$, where $M(X)$ denotes a fuzzy subset of U.

In figure 3 we can see an example of the linguistic variable "Height" with corresponding linguistic term set $T(Height) = \{Very\ Low, Low, Medium, High, Very\ High\}$. We can see how the semantic rule associates each of the linguistic terms of X to its fuzzy subset $M(X)$ of U. It is clear that a crucial aspect that will determine the validity of a Fuzzy Logic approach is the determination of correct membership functions for the linguistic term set [9]. It is worth mentioning that the membership function attached to a given word depends on its contextual intended use: *height* is not the same thing for a man or for a tree [10]. Therefore, the first task in modelling vague concepts is to define the universe of discourse. The meaning of a word will be influenced by the number of other words used to cover the universe of discourse previously defined. Indeed, the concept *young* does not mean exactly the same if the remaining vocabulary includes only *mature* and *old*, or is richer and contains both *very young*, *mature*, *old* and *very old*. Lastly, concepts might mean different to different people [11].

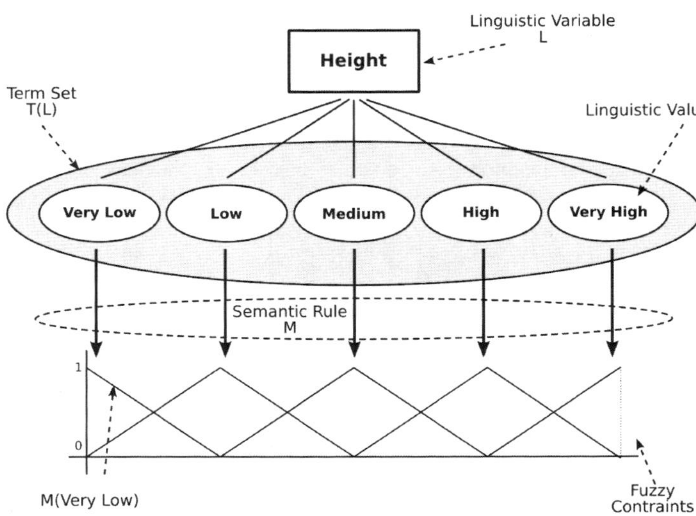

Fig. 3 Example of the linguistic variable "Height"

2.3 Basic Set-Theoretic Operations for Fuzzy Sets

The membership function is obviously the crucial component of a fuzzy set. It is therefore not surprising that operations with fuzzy sets are defined via their membership functions.

The union and intersection of classical sets are defined as follows:

$$x \in A \cap B \Leftrightarrow a \in A \wedge x \in B$$

$$x \in A \cup B \Leftrightarrow a \in A \vee x \in B$$

In terms of the characteristic function are defined as follows:

$$A \cap B = \{x | \delta_A(x) = 1 \wedge \delta_B(x) = 1\}$$

$$A \cup B = \{x | \delta_A(x) = 1 \vee \delta_B(x) = 1\}$$

Note that the use of the following characteristic functions

$$\delta_{A \cap B}(x) = \min\{\delta_A(x), \delta_B(x)\} \text{ and } \delta_{A \cup B}(x) = \max\{\delta_A(x), \delta_B(x)\}$$

for $A \cap B$ and $A \cup B$ respectively, is consistent with the above definitions. In fact, the min and max operators were originally suggested by Zadeh to model the intersection and union of fuzzy sets. Assuming that A and B are two fuzzy sets with membership functions μ_A and μ_B, the following operations can be defined on these sets [2, p. 310]:

Definition 2 (Intersection of Fuzzy Sets: min **operator)**

$$\mu_{A \cap B}(x) = \min\{\mu_A(x), \mu_B(x)\}$$

Definition 3 (Union of Fuzzy Sets: max **operator)**

$$\mu_{A \cup B}(x) = \max\{\mu_A(x), \mu_B(x)\}$$

Definition 4 (Complement of a Fuzzy Set)

$$\mu_{\overline{A}} = 1 - \mu_A(x)$$

Example 2. Consider the following fuzzy sets: A = "x approximately 11" and B = "x considerable larger than 10," characterized by the following membership functions

$$\mu_A(x) = \frac{1}{1 + (x - 11)^4}$$

and

$$\mu_B(x) = \begin{cases} 0 & \text{if } x \leq 10 \\ 1 - \frac{1}{1+(x-10)^2} & \text{if } x > 10. \end{cases}$$

respectively.

Figure 5 shows both fuzzy set graphics in the same coordinate axes, while the union and intersection of A and B, $A \cup B$ and $A \cap B$, are illustrated in figure 6:

The above are not the only possible ways to extend classical set theory consistently. Other operators have been also suggested. For example $\mu_{A \cap B}(x) = \mu_A(x) \cdot \mu_B(x)$, $\mu_{A \cup B}(x) = \mu_A(x) + \mu_B(x) - \mu_A(x) \cdot \mu_B(x)$ and $\mu_{\overline{A}} = 1 - \mu_A(x)$ are also consistent with classical set theory. These operators were also proposed by Zadeh in [2], and are special cases of a more general class of operators for the intersection and the union of fuzzy sets: *triangular norms* and *triangular conorms*. Another class of operators called *averaging operators* are also used to model connectives for fuzzy sets between t-norms and t-conorms. It is worth remarking that

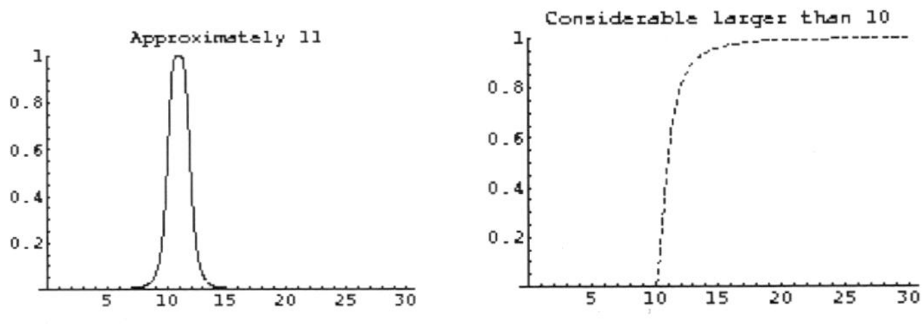

Fig. 4 Left: Fuzzy Set A; Right: Fuzzy Set B

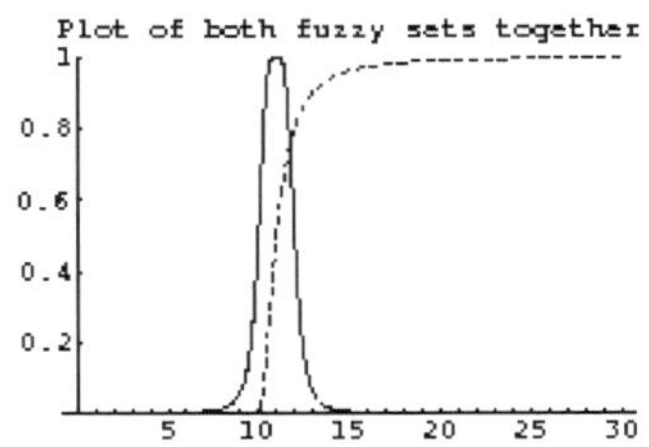

Fig. 5 Fuzzy sets A, B

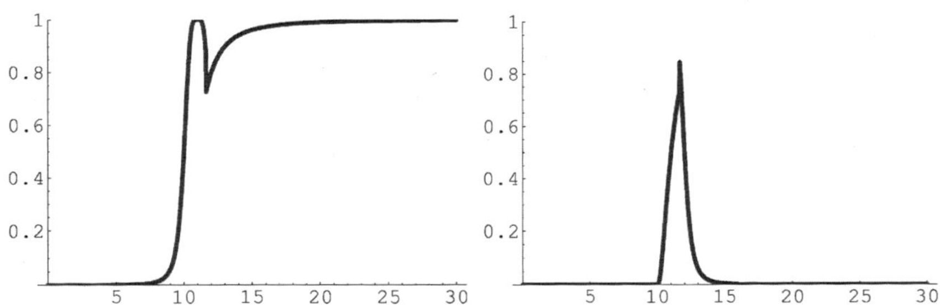

(a) $A \cup B$ (b) $A \cap B$

Fig. 6 Union and Intersection of Fuzzy Set A and B

when the intersection is interpreted as the "logical and" and the union as the "logical or", Bellman and Giertz proved in 1973 that under certain "reasonable" conditions both min and max operators should be used [12].

2.4 *Fuzzy Inference*

Fuzzy inference refers to the process leading to the evaluation of fuzzy rules of the form

$$\text{IF } x \text{ is } A \text{ AND } y \text{ is } B \text{ THEN } z \text{ is } C$$

where A, B and C are fuzzy sets defined on the universes of discourse X and Y respectively.

A fuzzy rule based system, known as fuzzy logic control (FLC), is one of the most well known applications of fuzzy logic. FLC is best utilised in complex ill defined processes that can be controlled by a skilled human operator without much knowledge of their underlying dynamics. The basic idea behind a FLC is to involve the 'expert experience' of a human operator in the design phase and express the input–output relationship of the process by a collection of 'if–then' rules as described above. The typical architecture diagram of a FLC is shown in figure 7 [13]:

- The *Knowledge Base* consists of the definition of the universe of discourse for each variable, the number of fuzzy set representations of input-output variables and the design of the membership functions to use.
- The *Fuzzification Unit* transforms crisp measured data (e.g. speed is $150km/h$) into suitable fuzzy sets (i.e. speed is too fast).
- The *Rule Base* stores the observed knowledge of the operation of the process which contains a set of 'if–then' rules.
- The *Inference Engine* is the main heart of a FLC and is capable of simulating human decision making by performing approximate reasoning.
- The *Defuzzification Unit* converts the fuzzy control action to a non-fuzzy action that can be accepted by the real world.

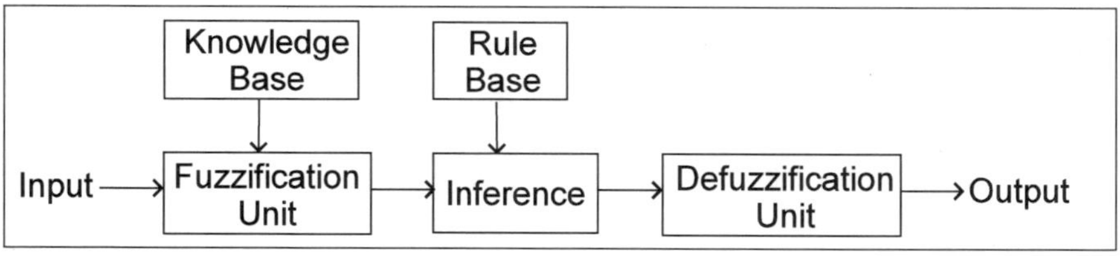

Fig. 7 FLC architecture diagram

The most commonly used fuzzy inference methods are:

- MAX-MIN fuzzy inference method
- MAX-DOT fuzzy inference method

For example, assume that we have a fuzzy control rule base with only two rules:

Rule 1 IF x is A_1 AND y is B_1 THEN z is C_1
Rule 2 IF x is A_2 AND y is B_2 THEN z is C_2

For inputs x_0 and y_0, the fire strengths of each rule are calculated as follows:

$$\alpha_1 = \min\{\mu_{A_1}(x_0), \mu_{B_1}(y_0)\}$$
$$\alpha_2 = \min\{\mu_{A_2}(x_0), \mu_{B_2}(y_0)\}$$

MAX–MIN Mandanis's minimum operator rule is used for fuzzy implication

Rule 1: $\mu_{R_1}(w) = \min\{\alpha_1, \mu_{C_1}(w)\}$
Rule 2: $\mu_{R_2}(w) = \min\{\alpha_2, \mu_{C_2}(w)\}$
Rule 1 OR Rule2: $\mu_{R_1 \vee R_2}(w) = \max\{\mu_{R_1}(w), \mu_{R_2}(w)\}$

MAX–DOT Larsen's product operation rule is used for fuzzy implication

Rule 1: $\alpha_1 = \mu_{A_1}(x_0) \cdot \mu_{B_1}(y_0)$
Rule 2: $\alpha_2 = \mu_{A_2}(x_0) \cdot \mu_{B_2}(y_0)$
Rule 1 OR Rule2: $\mu_{R_1 \vee R_2}(w) = \max\{\mu_{R_1}(w), \mu_{R_2}(w)\}$

In any case, as pointed out in [10] 'what is really important in practice is to correctly represent the pieces of knowledge provided by an expert and capture the meaning he intends to give to his own words.'

3 Application of Fuzzy Logic in Security

Computer Security concerns the protection of information and property from misuse by unauthorised people. Fuzzy logic can provide a good security analysis because it has the ability to take advantage of an operator's experience [14]. Another important aspect of fuzzy logic is the speed of response of the system. This is a key factor in network security [15]. Fuzzy logic can be used to predefine attacks and to store them. These stored attacks can in turn be used when a new attack occurs by comparing it with the previously stored ones. Fuzzy logic is currently used in computers to recognise assaults and compare them with attack signatures which have been stored in databases helping to determine whether security has been compromised [16].

Security threats can be detected by using a combination of fuzzy inference rules and linguistic variables without the need to use a clear mathematical formula which often takes a great effort to develop [17]. For example, a set of rules can be taken as conditions on which the result depends, such as:

If an input is *unknown* **And** behaviour is *unfamiliar* **Then** the risk is *very likely*

Because they account for partial degrees of membership, fuzzy systems can classify inputs. Most systems that have a security function will discover whether an input is normal or not. Fuzzy logic can take this on step further by further classifying abnormal inputs as being low, medium or very dangerous attacks. This can be achieved by applying a set of fuzzy rules to the security system [18]. In addition, fuzzy rules can easily be contracted to detect common security attacks by using if-then rules. This way, attacks can be classified according to their degree of dangerousness [19]. Fuzzy rules can be written by security administrators using their experience. The fuzzy expert can then verify that the rules cover all inputs [20]. Therefore, attacks can easily be detected with fuzzy systems based on sets of rules derived from the expert knowledge of the security administrator.

Computer viruses and malicious software can be detected by traditional antivirus software such as Kaspersky[1] and Norton Antivirus[2]. The problem with these programmes is that they can seek and detect only known virus patterns; in other words, unknown computer viruses cannot be detected putting the systems at risk. This has opened new axis of research including behaviour based systems. These systems can close the gaps between behaviour-based and knowledge-based system. Fuzzy logic systems such as the *fuzzy pattern recognition* can be used to detect unknown malicious software and computer viruses based on their behaviour [21].

3.1 Selected Fuzzy-Based Approaches to Intrusion Detection

Fuzzy systems have a number of aspects which make them suitable for intrusion detection [20]. Firstly, they have the ability to combine inputs taken from different sources. Secondly, they can cope with intrusions where some encroachments cannot be clearly determined. Thirdly, fuzzy systems have the ability to recognise the degree of membership. This can make alarms respond differently depending on the level of the attack (low-medium-high). In addition, fuzzy rule-based detection systems provide a perfect method to capture the imprecise tendency of occurrences in the computer network system [22]. Fuzzy logic is well suited for situations when the differences between normal and irregular classes are not specified. This is the case when fuzzy spaces enable an object to be a member of different classes at the same time [15].

Intrusion detection systems need to take account of a number of quantitative characteristics. So, in order to recognise high-level patterns, fuzzy set theory needs the ability to classify these quantitative characteristics. In addition, in quantitative characteristics, normal and abnormal cannot be directly separated. As a result, fuzzy systems can be used to distinguish between quantitative and non-quantitative characteristics based on fuzzy rules [18].

Anomalous behaviour can be detected by means of statistical analysis. However, normal patterns of statistical features can also be constructed automatically and more efficiently using fuzzy systems. These systems can then be utilised for anomaly detection [18]. The Fuzzy Intrusion Recognition Engine (FIRE) is an example of an intrusion detection system which is used to detect anomalous behaviour using fuzzy principles to recognize malevolent network actions [19]. FIRE uses autonomous fuzzy agents which work together to supervise the incoming data in the network. Depending on past abnormal behaviour which has been detected, an agent can detect whether these inputs are normal or not [23].

Classification and clustering algorithms based on fuzzy logic are efficient at dealing with malicious attacks. A fuzzy classifier can be used to solve intrusion detection problems by having two sets of rules, one for normal behaviour and another one for abnormal one [5]. This can be achieved by taking normal behaviour as the basis of the rules. Any activity which falls outside these norms can then be classified as

[1] `http://www.kaspersky.co.uk`
[2] `http://www.symantec.com/`

abnormal. Therefore, because of the ability of fuzzy logic to recognise partial membership, attacks can be detected as representing low, medium or high risk. Fuzzy clustering is particularly efficient at dealing with intrusion detection. This is specially the case with outlier attacks where a Fuzzy clustering method can be used to classify abnormal behaviour [14]. New inputs are compared with these clusters to detect whether they are abnormal or not.

A number of algorithms are used to improve intrusion detection systems using fuzzy logic. One such algorithm is the fuzzy C-means proposed by [20]. This algorithm is used with fuzzy rules to monitor network traffic, in order to detect malicious behaviour by checking the IP address of every single packet that is sent to the computer system. Other researchers have integrated fuzzy logic with other techniques to make IDS operate more efficiently. Fuzzy logic has been integrated with some data mining methods in [18] because these have the ability to extract patterns from large amounts of data in an automatic and adaptive manner. Consequently, fuzzy logic can be used on its own to detect attacks or combined with other machine learning techniques to achieve better results towards security problems.

4 Applications of Fuzzy Logic in Computer Forensics

Computer Forensics provides methods for the preservation, collection, validation, identification, analysis, interpretation, documentation and presentation of digital evidence derived from digital sources for the reconstruction of criminal activities or policy violations. Computer forensics should be improved to identify related events and unfamiliar characters of instructions. In computing forensics investigators are faced with many uncertainties. Fuzzy logic has proved to be an appropriate tool for forensics investigators providing more information related to these uncertainties [24].

An important operation in any computer forensics investigation is to analyse data related to exchanges of information between people, such as by email or chatting. The problem is that people often make mistakes (misspelled or use of the wrong word) when writing. This makes it more difficult for a system to analyse this data. A fuzzy engine can be used to solve this problem by relying on the selection of the degree of fuzziness. This can be achieved by using a fuzzy expert system which can take misspelled or wrong words and find relationships with similar words [25].

A fuzzy expert system has been used by [26] in network forensics. A set of components were used to achieve this system, as shown in Figure 8. First, a traffic analyser reads all incoming inputs and arranges them in groups. Fuzzification is then used to map crisp values into linguistic ones. The values which have been fuzzifieded are then compared with a set of fuzzy rules which have been determined by the security administrator. After comparing these values with the fuzzy rules, they then need to be defuzzified by the defuzzification component in order to convert them from output linguistic values into crisp output values. Finally, a forensics analyser is used to make a decision as to whether the values that have been defuzzified are attacks or not. If the forensics analyser component finds attacks, it will then

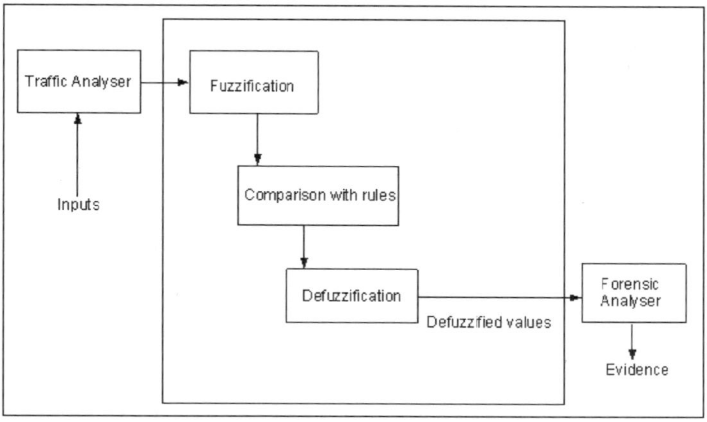

Fig. 8 Forensics using fuzzy expert system

assemble related data and store it as evidence. Thus, digital evidence can be gathered using a fuzzy expert system which will help investigators to identify the attacker.

One of the new technologies designed to help computer forensics investigations is Filter I [27]. This is a fuzzy logic filter that can be used to analyse data in files. The filter can be used in forensics to perform significant “analysis of windows swap file data, file slack data, and data associated with erased files”. It can also be used to find data in files associated with email, word processing, internet chat rooms, etc. [27, p.59]. In addition, this filter can be used in computer security as a penetration test tool to detect weaknesses of which the organisation is unaware.

When a forensic investigation takes place, the integrity leakage measures that are related to the gathering of evidence should be compulsory . It is very difficult to avoid integrity leakage while conducting a forensic investigation. This can be avoided by using the concept of fuzzy measures, which differ from traditional data analysis by having a great robustness and scalability. In addition, if fuzzy measures are not used during a forensic investigation, it is unlikely to have good experimental findings of wide scope that are appropriate to the real world [28]. Hence, fuzzy systems are valuable not just for gathering evidence, but also for avoiding systemic problems that might arise when performing a forensic investigation. This advantage of fuzzy systems makes them more suitable for forensic investigations than other techniques. Other techniques are often limited to collecting evidence when an attack has occurred while ignoring other data in the system. This ignored data might be more important than what has been lost.

5 Conclusion

This chapter has presented an introduction to the field of Fuzzy Logic together with some examples of its applications in computer security and computing forensics.

Fuzzy logic is a human-like reasoning approach that resembles the ambiguity of human daily life. It has the ability to describe core aspects that may not be apparent; that is, it can provide precise outcomes by its ability to assess degrees of membership and linguistic variables. Security in itself is imprecise, and therefore a fuzzy system provide a sound approach towards solving problems related to computer security and computer forensics. As can be seen in this chapter, there are a number of advantages to using fuzzy systems in security. They include defining attacks by applying a set of fuzzy rules and exploiting the associated speed of response. A vast amount of the information required on computer forensic investigations is uncertain. These include unclear file types, bad handwriting, and so on. Fuzzy systems have shown a great ability in dealing with these kinds of ambiguous cases.

References

1. Zadeh, L.A.: Outline of a new approach to the analysis of complex system and decision processes. IEEE Transactions on Systems Man and Cybernetics 3, 28–44 (1973)
2. Zadeh, L.A.: Fuzzy sets. Information and Control 8, 338–353 (1965)
3. Negnevitsky, M.: Artificial Intelligence: A guide to intelligent systems. Pearson Education Limited, Essex (2005)
4. Chen, G., Pham, T.: Introduction to fuzzy sets, fuzzy logic, and fuzzy control systems. CRC Press, Washington (2001)
5. Klir, G., Yuan, B.: Fuzzy sets and fuzzy logic: Theory and Applications. Prentice Hall PTR, New Jersey (1995)
6. Zadeh, L.A.: The concept of a linguistic variable and its applications to approximate reasoning, Part I. Information Sciences 8, 199–249 (1975)
7. Zadeh, L.A.: The concept of a linguistic variable and its applications to approximate reasoning, Part II. Information Sciences 8, 301–357 (1975)
8. Zadeh, L.A.: The concept of a linguistic variable and its applications to approximate reasoning, Part III. Information Sciences 9, 43–80 (1975)
9. Herrera, F., Alonso, S., Chiclana, F., Herrera-Viedma, E.: Computing with Words in Decision Making: Foundations. Trends and Prospects Fuzzy Optimization and Decision Making 8, 337–364 (2009)
10. Dubois, D., Ostasiewicz, W., Prade, H.: Fuzzy sets: history and basic notions. In: Dubois, D., Prade, H. (eds.) Fundamentals of Fuzzy Sets. Kluwer Academic Publishers (2000)
11. Mendel, J.M.: Uncertain Rule-Based Fuzzy Logic Systems: Introduction and New Directions. Prentice Hall, Upper Saddle River (2001)
12. Belman, R., Giertz, M.: On the analytic formalism of the theory of fuzzy sets. Information Science 5, 149–156 (1973)
13. Berenji, H.R.: Fuzzy logic controllers. In: Yager, R.R., Zadeh, L.A. (eds.) An Introduction to Fuzzy Logic Applications in Intelligent Systems. Kluwer Academic Publishers (1992)
14. Heydeman, J., Reijntjes, R., Babuska, R., Kaymak, U., Van Nauta Lemke, H.: Fuzzy Logic Based Security Assessment of Power Networks. Intelligent Systems Applications to Power Systems 2(9), 405–409 (1996)
15. Gomez, J., Dasgupta, D.: Evolving Fuzzy Classifiers for Intrusion Detection. In: Proceedings of 2002 IEEE Workshop on Information Assurance. United States Military Academy, IEEE, West Point, NY (2002)

16. Miller, S.: Wi-Fi Security. McGraw-Hill, New York (2003)
17. McGill, W., Ayyub, B.: A Multicriteria Security System Performance Assessment Using Fuzzy Logic. The Journal of Defense Modeling and Simulation: Applications, Methodology, Technology 4(4), 484–504 (2007)
18. Luo, J.: Integrating Fuzzy Logic with Data Mining Methods for Intrusion Detection. Unpublished MSc. thesis. Mississippi State University (1999)
19. Dickerson, J.E., Dickerson, J.A.: Fuzzy Network Profiling for Intrusion Detection. In: 19th International Conference of the North American Fuzzy Information Processing Society, pp. 301–306. IEEE, Atlanta (2000)
20. Dickerson, J.E., Juslin, J., Koukousoula, O., Dickerson, J.A.: Fuzzy Intrusion Detection. In: Joint 9th IFSA World Congress and 20th NAFIPS International Conference, 2001, Vancouver, pp. 1506–1510. IEEE, Piscataway (2001)
21. Zhang, B., Yin, J., Hag, J.: Using Fuzzy Pattern Recognition to Detect Unknown Malicious Executables Code. In: Wang, L., Jin, Y. (eds.) Fuzzy Systems and Knowledge Discovery, pp. 629–634. Springer, Heidelberg (2005)
22. Yao, L., ZhiTang, L., Shuyu, L.: A Fuzzy Anomaly Detection Algorithm for IPv6. In: Second IEEE International Conference on Semantics, Knowledge and Grid, SKG 2006, Guilin, Guangxi, China, pp. 67–70 (2006)
23. Xin, J., Dickerson, J.E., Dickerson, J.A.: Fuzzy feature extraction and visualization for intrusion detection. In: 12th IEEE International Conference on Fuzzy Systems, 2003, pp. 1249–1254. IEEE, St. Louis. MO. (2003)
24. Meyers, M.: Computer Forensics: Towards Creating A Certification Framework. Unpublished MSc. thesis. Purdue University (2005)
25. Johansson, C.: Computer Forensic Text Analysis with Open Source Software. Unpublished MSc. thesis. Blekinge Institute of Technology (2003)
26. Kim, J., Kim, D., Noh, B.: A Fuzzy Logic Based Expert System as a Network Forensics. In: IEEE International Conference on Fuzzy Systems 2004, Budapest, pp. 879–884 (2004)
27. Wai, H.: Research in Computer Forensics. Unpublished MSc. thesis. Naval Postgraduate School (2002)
28. Pan, L., Savoldi, A., Gubian, B., Batten, L.: Measure of Integrity Leakage in Live Forensic Context. In: 2008 IEEE International Conference on Intelligent Information Hiding and Multimedia Signal Processing Harbin, China, pp. 534–537 (2008)

A Topological Study of Chaotic Iterations Application to Hash Functions

Christophe Guyeux and Jacques M. Bahi

Abstract. Chaotic iterations, a tool formerly used in distributed computing, has recently revealed various interesting properties of disorder leading to its use in the computer science security field. In this paper, a comprehensive study of its topological behavior is proposed. It is stated that, in addition to being chaotic as defined in the Devaney's formulation, this tool possesses the property of topological mixing. Additionally, its level of sensibility, expansivity, and topological entropy are evaluated. All of these properties lead to a complete unpredictable behavior for the chaotic iterations. As it only manipulates binary digits or integers, we show that it is possible to use it to produce truly chaotic computer programs. As an application example, a truly chaotic hash function is proposed in two versions. In the second version, an artificial neural network is used, which can be stated as chaotic according to Devaney.

Keywords: Topological chaos, Chaotic iterations, Hash functions, Neural Networks.

1 Introduction

Chaotic iterations (CIs) were formerly a way to formalize distributed algorithms through mathematical tools [9]. By using these CIs, it was thus possible to study the convergence of synchronous or asynchronous programs over parallel, distributed, P2P, grid, or GPU platforms, in a view to solve linear and non-linear systems. We have proven at the IEEE World Congress on Computational Intelligence (WCCI'10) that CIs can behave chaotically, as it is defined by Devaney [5]. These proofs have

Christophe Guyeux · Jacques M. Bahi
Computer Science Laboratory LIFC, University of Franche-Comté
16, route de Gray - 25030 Besançon, France
Phone: +33 381666948
e-mail: {christophe.guyeux, jacques.bahi}@univ-fcomte.fr

D.A. Elizondo et al. (Eds.): Computational Intelligence for Privacy and Security, SCI 394, pp. 51–73.
springerlink.com

been improved and more detailed in [4]. In this paper, which is an extension of [5, 4], we notably enlarge the theoretical study of CIs, among other things by computing its topological entropy, to obtain a comprehensive evaluation of its topological behavior. This study leads us to the conclusion that the chaos of CIs is very intense and constitutes a useful tool to be used in the computer science security field.

Chaos in information security fields as digital watermarking [10, 11], hash functions [24, 26], or pseudo-random number generators, is often disputed. This is due to the fact that this use is almost always based on the conception of algorithms that only include "somewhere" some well-known chaotic real functions like logistic, tent, or Arnold's cat maps, to obtain a program supposed to express these chaotic properties [14, 25, 22]. However, using such functions with other "obvious" parameters does not guarantee that the whole algorithm still remains chaotic. Such an assumption should at least be discussed. Moreover, even if the algorithm obtained by the inclusion of chaotic maps is itself proven to be chaotic, its implementation on a machine can lead to the fact that this chaotic nature is lost. This is due to the finite cardinality of the machine numbers set.

In this paper, as in [5, 4], we do not simply integrate chaotic maps into algorithms hoping that the result remains chaotic. We conceive algorithms for computer security that we have mathematically proven to be chaotic, as it is defined in the Devaney's theory. We raise the question of their implementation, proving in doing so that it is possible to design both a chaotic algorithm and its associated chaotic computer program. The chaos theory we consider is taken from the mathematical topology. It encompasses the well-known Devaney's definition of chaos and the notions of expansivity, topological entropy, and topological mixing. These notions of unpredictability are the most established ones into the mathematical theory of chaos. Our fundamental study is motivated by the desire to produce chaotic programs in the area of information security.

The paper begins by introducing the theoretical foundation of this approach. On the one hand we recall the definition of Devaney's topological chaos and on the other hand the definition of discrete chaotic iterations. Although these definitions are distinct from each other, we establish a link between them by giving conditions under which chaotic discrete iterations generate a Devaney's topological chaos. This study is deepened by giving some qualitative and quantitative evaluations of the disorder generated by chaotic iterations. These evaluations are not present in [4, 5]. We will focus in this paper on the notions of expansivity, topological mixing, and topological entropy. The proofs that the considered space is separated and compact have never been published before. Furthermore, the evaluation of the topological entropy is completely new. Then, because chaotic iterations are very suited for computer programming, this link allows us to generate programs in the computer science field that behave chaotically. This link was formerly presented in [5, 4] with some errors that are corrected here.

After having studied the theoretical aspects of our approach we focus on the practical ones. The important question is how to preserve the topological chaos properties in a set of a finite number of states. This question is answered in Section 5, by manipulating only integers and considering the use of new data at each iteration.

The general algorithm based on our approach, formerly presented in [5], is now explained in detail in Section 6. It is rewritten in the next section as an artificial neural network that can compute hash values while behaving chaotically. This new application is completely new and has never been published before.

The remainder of this paper is organized in the following way. In Section 2, the definitions of Devaney's chaos and discrete chaotic iterations are recalled. A link between these two notions is established and sufficient conditions to obtain Devaney's topological chaos from discrete chaotic iterations are given in Section 3. The chaotic behavior of CIs is deepened in Section 4, by studying some qualitative and quantitative properties of disorder. In Section 5, the question on how to preserve these chaotic properties into computers is answered. Then in Section 6 the general hash function scheme is given and illustrated, whereas in Section 7 it is applied to produce an artificial neural network able to hash some values in a chaotic manner. The paper ends by a conclusion section in which our contribution is summarized and planned future work is discussed.

2 Basic Recalls

This section is devoted to basic definitions and terminologies in the field of topological chaos and in the one of chaotic iterations.

2.1 Devaney's Chaotic Dynamical Systems

In the sequel S^n denotes the nth term of a sequence S and V_i denotes the ith component of a vector V. $f^k = f \circ ... \circ f$ denotes the kth composition of a function f. Finally, the following notation is used: $[\![1;N]\!] = \{1,2,\dots,N\}$.

Consider a topological space $(\mathcal{X},\tau)$ and a continuous function $f : \mathcal{X} \to \mathcal{X}$.

Definition 1. f is said to be *topologically transitive* if, for any pair of open sets $U,V \subset \mathcal{X}$, there exists $k > 0$ such that $f^k(U) \cap V \neq \varnothing$.

Definition 2. An element x is a *periodic point* for f of period $n \in \mathbb{N}^*$ if $f^n(x) = x$.

Definition 3. f is said to be *regular* on $(\mathcal{X},\tau)$ if the set of periodic points for f is dense in $\mathcal{X}$: for any point x in $\mathcal{X}$, any neighborhood of x contains at least one periodic point (without necessarily the same period).

Definition 4. f is said to be *chaotic* on $(\mathcal{X},\tau)$ if f is regular and topologically transitive.

The chaos property is strongly linked to the notion of "sensitivity", defined on a metric space $(\mathcal{X},d)$ by:

Definition 5. f has *sensitive dependence on initial conditions* if there exists $\delta > 0$ such that, for any $x \in \mathcal{X}$ and any neighborhood V of x, there exist $y \in V$ and $n > 0$ such that $d(f^n(x), f^n(y)) > \delta$.

δ is called the *constant of sensitivity* of f.

Indeed, Banks *et al.* have proven in [7] that when f is chaotic and $(\mathscr{X},d)$ is a metric space, then f has the property of sensitive dependence on initial conditions (this property was formerly an element of the definition of chaos). To sum up, quoting Devaney in [13], a chaotic dynamical system "is unpredictable because of the sensitive dependence on initial conditions. It cannot be broken down or simplified into two subsystems which do not interact because of topological transitivity. And in the midst of this random behavior, we nevertheless have an element of regularity". Fundamentally different behaviors are consequently possible and occur in an unpredictable way.

2.2 Chaotic Iterations

Let us consider a *system* with a finite number $\mathsf{N} \in \mathbb{N}^*$ of elements (or *cells*), so that each cell has a Boolean *state*. Having N Boolean values for these cells leads to the definition of a particular *state of the system*. A sequence which elements belong to $[\![1;\mathsf{N}]\!]$ is called a *strategy*. The set of all strategies is denoted by $\mathbb{S}$.

Definition 6. The set $\mathbb{B}$ denoting $\{0,1\}$, let $f : \mathbb{B}^{\mathsf{N}} \longrightarrow \mathbb{B}^{\mathsf{N}}$ be a function and $S \in \mathbb{S}$ be a strategy. The so-called *chaotic iterations* are defined by $x^0 \in \mathbb{B}^{\mathsf{N}}$ and

$$\forall n \in \mathbb{N}^*, \forall i \in [\![1;\mathsf{N}]\!], x_i^n = \begin{cases} x_i^{n-1} & \text{if } S^n \neq i \\ \left(f(x^{n-1})\right)_{S^n} & \text{if } S^n = i. \end{cases}$$

In other words, at the nth iteration, only the S^n-th cell is "iterated". Note that in a more general formulation, S^n can be a subset of components and $\left(f(x^{n-1})\right)_{S^n}$ can be replaced by $\left(f(x^k)\right)_{S^n}$, where $k < n$, describing for example, delays transmission [20, 15]. Finally, let us remark that the term "chaotic", in the name of these iterations, has *a priori* no link with the mathematical theory of chaos, recalled above.

3 Chaotic Iterations as Devaney'S Chaos

In this section is proven that chaotic iterations are a particular case of topological chaos, as it is defined in the Devaney's formulation'.

3.1 The New Topological Space

In this section we define a suitable metric space where chaotic iterations are continuous.

3.1.1 Defining the Iteration Function and the Phase Space

Let δ be the *discrete Boolean metric*, $\delta(x,y) = 0 \Leftrightarrow x = y$. Given a function f, define the function:

$$\begin{array}{rcl} F_f : [\![1;\mathsf{N}]\!] \times \mathbb{B}^{\mathsf{N}} & \longrightarrow & \mathbb{B}^{\mathsf{N}} \\ (k,E) & \longmapsto & \left(E_j.\delta(k,j) + f(E)_k.\overline{\delta(k,j)}\right)_{j \in [\![1;\mathsf{N}]\!]} \end{array},$$

where + and . are the Boolean addition and product operations. Consider the phase space:

$$\mathscr{X} = [\![1;\mathsf{N}]\!]^{\mathbb{N}} \times \mathbb{B}^{\mathsf{N}},$$

and the map defined on $\mathscr{X}$:

$$G_f(S,E) = \left(\sigma(S), F_f(i(S),E)\right), \tag{1}$$

where σ is the *shift* function defined by $\sigma(S^n)_{n\in\mathbb{N}} \in \mathbb{S} \longrightarrow (S^{n+1})_{n\in\mathbb{N}} \in \mathbb{S}$ and i is the *initial function* $i : (S^n)_{n\in\mathbb{N}} \in \mathbb{S} \longrightarrow S^0 \in [\![1;\mathsf{N}]\!]$. Then the chaotic iterations defined in (2.2) can be described by the following iterations:

$$\begin{cases} X^0 \in \mathscr{X} \\ X^{k+1} = G_f(X^k). \end{cases}$$

With this formulation, a shift function appears as a component of chaotic iterations. The shift function is a famous example of a chaotic map [13] but its presence is not sufficient enough to claim G_f as chaotic. In the rest of this section we rigorously prove that under some hypotheses, chaotic iterations generate topological chaos. Furthermore, due to the suitability of chaotic iterations for computer programming [3, 23, 2] we also have proven that this is true in the computer science field.

3.1.2 Cardinality of $\mathscr{X}$

By comparing $\mathbb{S}$ and $\mathbb{R}$, we have the following result.

Theorem 1. *The phase space $\mathscr{X}$ has, at least, the cardinality of the continuum.*

Proof. Let φ be the map which transforms a strategy into the binary representation of an element in $[0,1[$, as follows. If the nth term of the strategy is 0, then the nth associated digit is 0. If this n^{th} term is not equal to 0, then the associated digit is 1. With this construction, $\varphi : [\![1;\mathsf{N}]\!]^{\mathbb{N}} \longrightarrow [0,1]$ is onto. But $]0,1[$ is isomorphic to $\mathbb{R}$ ($x \in]0,1[\mapsto tan(\pi(x-\frac{1}{2}))$ is an isomorphism), so the cardinality of $[\![1;\mathsf{N}]\!]^{\mathbb{N}}$ is greater or equal to the cardinality of $\mathbb{R}$. As a consequence, the cardinality of the Cartesian product $\mathscr{X} = [\![1;\mathsf{N}]\!]^{\mathbb{N}} \times \mathbb{B}^{\mathsf{N}}$ is greater or equal to the cardinality of $\mathbb{R}$.

Remark 1. This result is independent from the number of components of the system.

3.1.3 A New Distance

Let us define a new distance between two points $X = (S,E), Y = (\check{S},\check{E}) \in \mathscr{X}$ by

$$d(X,Y) = d_e(E,\check{E}) + d_s(S,\check{S}),$$

where

$$\begin{cases} d_e(E,\check{E}) = \sum_{k=1}^{\mathsf{N}} \delta(E_k,\check{E}_k), \\ d_s(S,\check{S}) = \frac{9}{\mathsf{N}} \sum_{k=1}^{\infty} \frac{|S^k - \check{S}^k|}{10^k}. \end{cases}$$

This new distance has been introduced in [5] to satisfy the following requirements.

- When the number of different cells between two systems is increasing, then their distance should increase too.
- In addition, if two systems present the same cells and their respective strategies start with the same terms, then the distance between these two points must be small because the evolution of the two systems will be the same for a while. Indeed, the two dynamical systems start with the same initial condition, use the same update function, and as strategies are the same for a while, then components that are updated are the same too.

The distance presented above follows these recommendations. Indeed, if the floor value $\lfloor d(X,Y) \rfloor$ is equal to n, then the systems $E,\check{E}$ differ in n cells. In addition, $d(X,Y) - \lfloor d(X,Y) \rfloor$ is a measure of the differences between strategies S and $\check{S}$. More precisely, this floating part is less than 10^{-k} if and only if the first k terms of the two strategies are equal. Moreover, if the kth digit is nonzero, then the kth terms of the two strategies are different.

3.1.4 Continuity of the Iteration Function

To prove that chaotic iterations are an example of topological chaos in the sense of Devaney [13], G_f must be continuous in the metric space $(\mathscr{X},d)$.

Theorem 2. *G_f is a continuous function.*

Proof. We use the sequential continuity. Let $(S^n,E^n)_{n\in\mathbb{N}}$ be a sequence of the phase space $\mathscr{X}$, which converges to (S,E). We will prove that $\left(G_f(S^n,E^n)\right)_{n\in\mathbb{N}}$ converges to $\left(G_f(S,E)\right)$. Let us recall that for all n, S^n is a strategy, thus, we consider a sequence of strategies (*i.e.*, a sequence of sequences).
As $d((S^n,E^n);(S,E))$ converges to 0, each distance $d_e(E^n,E)$ and $d_s(S^n,S)$ converges to 0. But $d_e(E^n,E)$ is an integer, so $\exists n_0 \in \mathbb{N}$, $d_e(E^n,E) = 0$ for any $n \geqslant n_0$.
In other words, there exists a threshold $n_0 \in \mathbb{N}$ after which no cell will change its state: $\exists n_0 \in \mathbb{N}, n \geqslant n_0 \Rightarrow E^n = E$.

In addition, $d_s(S^n,S) \longrightarrow 0$, so $\exists n_1 \in \mathbb{N}, d_s(S^n,S) < 10^{-1}$ for all indexes greater than or equal to n_1. This means that for $n \geqslant n_1$, all the S^n have the same first term, which is S^0: $\forall n \geqslant n_1, S_0^n = S_0$.

Thus, after the $max(n_0,n_1)^{th}$ term, states of E^n and E are identical and strategies S^n and S start with the same first term.
Consequently, states of $G_f(S^n,E^n)$ and $G_f(S,E)$ are equal, so, after the $max(n_0,n_1)^{th}$ term, the distance d between these two points is strictly less than 1.

We now prove that the distance between $\left(G_f(S^n,E^n)\right)$ and $\left(G_f(S,E)\right)$ is convergent to 0. Let $\varepsilon > 0$.

- If $\varepsilon \geqslant 1$, we see that distance between $\left(G_f(S^n,E^n)\right)$ and $\left(G_f(S,E)\right)$ is strictly less than 1 after the $max(n_0,n_1)^{th}$ term (same state).
- If $\varepsilon < 1$, then $\exists k \in \mathbb{N}, 10^{-k} \geqslant \varepsilon > 10^{-(k+1)}$. But $d_s(S^n,S)$ converges to 0, so

$$\exists n_2 \in \mathbb{N}, \forall n \geqslant n_2, d_s(S^n,S) < 10^{-(k+2)},$$

thus after n_2, the $k+2$ first terms of S^n and S are equal.

As a consequence, the $k+1$ first entries of the strategies of $G_f(S^n,E^n)$ and $G_f(S,E)$ are the same (G_f is a shift of strategies) and due to the definition of d_s, the floating part of the distance between (S^n,E^n) and (S,E) is strictly less than $10^{-(k+1)} \leqslant \varepsilon$.

In conclusion,

$$\forall \varepsilon > 0, \exists N_0 = max(n_0,n_1,n_2) \in \mathbb{N}, \forall n \geqslant N_0, d\left(G_f(S^n,E^n);G_f(S,E)\right) \leqslant \varepsilon.$$

G_f is consequently continuous.

In this section, we proved that chaotic iterations can be modeled as a dynamical system in a topological space. In the next section, we show that some chaotic iterations behave chaotically, as defined by Devaney's theory.

3.2 Discrete Chaotic Iterations as Topological Chaos

To prove that we are in the framework of Devaney's topological chaos, we have to find a Boolean function f such that G_f satisfies the regularity, transitivity, and sensitivity conditions. We will prove that the vectorial logical negation

$$f_0(x_1,\ldots,x_\mathsf{N}) = (\overline{x_1},\ldots,\overline{x_\mathsf{N}}) \qquad (2)$$

is a suitable function.

3.2.1 Regularity

Firstly, let us prove that,

Theorem 3. *Periodic points of G_{f_0} are dense in $\mathscr{X}$.*

Proof. Let $(\check{S},\check{E}) \in \mathscr{X}$ and $\varepsilon > 0$. We are looking for a periodic point $(\widetilde{S},\widetilde{E})$ satisfying $d((\check{S},\check{E});(\widetilde{S},\widetilde{E})) < \varepsilon$. As ε can be strictly lesser than 1, we must choose $\widetilde{E} = \check{E}$. Let us define $k_0(\varepsilon) = \lfloor log_{10}(\varepsilon) \rfloor + 1$ and consider the set

$$\mathscr{S}_{\check{S},k_0(\varepsilon)} = \left\{ S \in \mathbb{S} / S^k = \check{S}^k, \forall k \leqslant k_0(\varepsilon) \right\}.$$

Then, $\forall S \in \mathscr{S}_{\check{S},k_0(\varepsilon)}, d((S,\check{E});(\check{S},\check{E})) < \varepsilon$. It remains to choose $\widetilde{S} \in \mathscr{S}_{\check{S},k_0(\varepsilon)}$ such that $(\widetilde{S},\widetilde{E}) = (\widetilde{S},\check{E})$ is a periodic point for G_{f_0}. Let

$$\mathscr{I} = \left\{ i \in \{1,...,\mathsf{N}\} / E_i \neq \check{E}_i, \text{ where } (S,E) = G_{f_0}^{k_0}(\check{S},\check{E}) \right\},$$

$i_0 = card(\mathscr{I})$, and $j_1 < j_2 < ... < j_{i_0}$ the elements of $\mathscr{I}$. Then, $\widetilde{S} \in \mathscr{S}_{\check{S},k_0(\varepsilon)}$ defined by

- $\widetilde{S}^k = \check{S}^k$, if $k \leqslant k_0(\varepsilon)$,
- $\widetilde{S}^k = j_{k-k_0(\varepsilon)}$, if $k \in \{k_0(\varepsilon)+1, k_0(\varepsilon)+2, ..., k_0(\varepsilon)+i_0\}$,
- and $\widetilde{S}^k = \widetilde{S}^j$, where $j \leqslant k_0(\varepsilon)+i_0$ is satisfying $j \equiv k \pmod{k_0(\varepsilon)+i_0}$, if $k > k_0(\varepsilon)+i_0$,

is such that $(\widetilde{S},\widetilde{E})$ is a periodic point (of period $k_0(\varepsilon)+i_0$), which is $\varepsilon-$close to $(\check{S},\check{E})$.
As a conclusion, $(\mathscr{X}, G_{f_0})$ is regular.

3.2.2 Transitivity

Regarding the transitivity property of G_{f_0}, we can show that,

Theorem 4. *$(\mathscr{X}, G_{f_0})$ is topologically transitive.*

Proof. Let us define $\mathscr{E} : \mathscr{X} \to \mathbb{B}^{\mathsf{N}}$, such that $\mathscr{E}(S,E) = E$. Let $\mathscr{B}_A = \mathscr{B}(X_A, r_A)$ and $\mathscr{B}_B = \mathscr{B}(X_B, r_B)$ be two open balls of $\mathscr{X}$, with $X_A = (S_A, E_A)$ and $X_B = (S_B, E_B)$. We are looking for $\widetilde{X} = (\widetilde{S},\widetilde{E})$ in $\mathscr{B}_A$ such that $\exists n_0 \in \mathbb{N}, G_{f_0}^{n_0}(\widetilde{X}) \in \mathscr{B}_B$.
$\widetilde{X}$ must be in $\mathscr{B}_A$ and r_A can be strictly lesser than 1, so $\widetilde{E} = E_A$. Let $k_0 = \lfloor \log_{10}(r_A) + 1 \rfloor$. Then $\forall S \in \mathbb{S}$, if $S^k = S_A^k, \forall k \leqslant k_0$, then $(S,\widetilde{E}) \in \mathscr{B}_A$. Let $(\check{S},\check{E})$ be equal to $G_{f_0}^{k_0}(S_A, E_A)$ and $c_1, ..., c_{k_1}$ denote the elements of the set $\{i \in [\![1,\mathsf{N}]\!] / \check{E}_i \neq \mathscr{E}(X_B)_i\}$. So any point X of the set

$$\{(S,E_A) \in \mathscr{X} / \forall k \leqslant k_0, S^k = S_A^k \text{ and } \forall k \in [\![1,k_1]\!], S^{k_0+k} = c_k\}$$

is satisfying $X \in \mathscr{B}_A$ and $\mathscr{E}\left(G_{f_0}^{k_0+k_1}(X)\right) = E_B$. Lastly, let k_2 be $\lfloor \log_{10}(r_B) \rfloor + 1$. Then $\widetilde{X} = (\widetilde{S},\widetilde{E}) \in \mathscr{X}$ defined by:

1. $\widetilde{X} = E_A$,
2. $\forall k \leqslant k_0, \widetilde{S}^k = S_A^k$,
3. $\forall k \in [\![1,k_1]\!], \widetilde{S}^{k_0+k} = c_k$,
4. $\forall k \in \mathbb{N}^*, \widetilde{S}^{k_0+k_1+k} = S_B^k$,

is such that $\widetilde{X} \in \mathscr{B}_A$ and $G_{f_0}^{k_0+k_1}(\widetilde{X}) \in \mathscr{B}_B$. This fact concludes the proof of the theorem.

3.2.3 Devaney's Chaos

In conclusion, $(\mathcal{X}, G_{f_0})$ is topologically transitive and regular. Then we have the following result:

Theorem 5. *G_{f_0} is a chaotic map on $(\mathcal{X}, d)$ in the sense of Devaney.*

We have proven that the set $\mathcal{C}$ of the iterate functions f so that $(\mathcal{X}, G_f)$ is chaotic (according to the definition of Devaney), is a nonempty set. In future work, we will deepen the study of $\mathcal{C}$, among other things, by computing its cardinality and characterizing this set.

4 Topological Properties of Chaotic Iterations

In this section, some qualitative and quantitative topological properties for chaotic iterations with G_{f_0} will be studied in detail. These properties reinforce the chaotic behavior of the system.

4.1 Topological Mixing

The topological mixing is a strong version of transitivity:

Definition 7. A discrete dynamical system is said to be *topologically mixing* if and only if, for any couple of disjoint open set $U, V \neq \varnothing$, $n_0 \in \mathbb{N}$ can be found so that $\forall n \geqslant n_0, f^n(U) \cap V \neq \varnothing$.

We have the result [16],

Theorem 6. *$(\mathcal{X}, G_{f_0})$ is topologically mixing.*

This result is an immediate consequence of the lemma below.

Lemma 1. *For any open ball B of $\mathcal{X}$, an index n can be found such that $G_{f_0}^n(B) = \mathcal{X}$.*

Proof. Let $B = B((E,S), \varepsilon)$ be an open ball, which the radius can be considered as strictly less than 1. The elements of B all have the same state E and are such that an integer $k\,(= -\log_{10}(\varepsilon))$ satisfies:

- all the strategies of B have the same k first terms,
- after the index k, all values are possible.

Then, after k iterations, the new state of the system is $G_{f_0}^k(E,S)_1$ and all the strategies are possibles (any point of the form $(G_{f_0}^k(E,S)_1, \hat{S})$, with any $\hat{S} \in \mathbb{S}$, is reachable from B).

Let $(E', S') \in \mathcal{X}$. We will prove that any point of $\mathcal{X}$ is reachable from B.

Indeed, let s_i be the list of the different cells between $G_{f_0}^k(E,S)_1$ and E', and $|s|$ its size. The point $(\check{E}, \check{S})$ of B defined by:

- $\check{E} = E$,
- $\check{S}^i = S^i, \forall i \leqslant k$,
- $\check{S}^{k+i} = s_i, \forall i \leqslant |s|$,
- $\forall i \in \mathbb{N}, S^{k+|s|+i} = S'^i$.

is such that $G_{f_0}^{k+|s|}(\check{E},\check{S}) = (E',S')$. This conclude the proof of the lemma.

4.2 *Quantitative Measures*

4.2.1 General Definitions

In Section 3.2.3 we have proven that discrete chaotic iterations produce a topological chaos by checking two qualitative properties, namely transitivity and regularity. This mathematical framework offers tools to measure this chaos quantitatively.
The first of these measures is the constant of sensitivity defined in Definition 5. Intuitively, a function f having a constant sensitivity equal to δ implies that there exists points arbitrarily close to any point x, which eventually separate from x by at least δ under some iterations of f. This induces that an arbitrarily small error on an initial condition *might be* magnified upon iterations of f.

Another important tool is defined below.

Definition 8. A function f is said to have the property of *expansivity* if

$$\exists \varepsilon > 0, \forall x \neq y, \exists n \in \mathbb{N}, d(f^n(x), f^n(y)) \geqslant \varepsilon.$$

Then, ε is the *constant of expansivity* of f. We also say that f is ε-expansive.

A function f has a constant of expansivity equal to ε if an arbitrarily small error on any initial condition is *always* magnified until ε.

4.2.2 Sensitivity

The sensitive dependence on the initial conditions has been shown as a consequence of the regularity and the transitivity of chaotic iterations. However, in the set of machine numbers, we have shown in [5] that the notion of regularity must be redefined. This is the reason why this sensitivity should be proven without using the result of Banks [7], to be sure that this dependence is preserved in practical use of chaotic iterations.

In addition, the constant of sensitivity will be obtained during this proof.

Theorem 7. *$(\mathcal{X}, G_{f_0})$ has sensitive dependence on initial conditions and its constant of sensitivity is equal to* $\mathsf{N}-1$.

Proof. Let $\check{X} = (\check{S},\check{E}) \in \mathcal{X}$. We are looking for $\widetilde{X} = (\widetilde{S},\widetilde{E}) \in \mathcal{X}$ such that $d(\check{X},\widetilde{X}) \leqslant \delta$ and $\exists n_0 \in \mathbb{N}$, $d\left(G_{f_0}^{n_0}(\check{X}); G_{f_0}^{n_0}(\widetilde{X})\right) \geqslant \mathsf{N}-1$. Let k_0 be $\lfloor \log_{10}(\delta) \rfloor + 1$. So, if $S \in \{S \in \mathbb{S} / \forall k \leqslant k_0, S^k = \check{S}^k\}$, then $d\left((S,\check{E}),(\check{S},\check{E})\right) \leqslant \delta$.

Let $\mathscr{J} = \left\{ i \in [\![1,\mathsf{N}]\!] / \ \mathscr{E}\left(G_{f_0}^{k_0}(\check{S},\check{E})\right)_i = \mathscr{E}\left(G_{f_0}^{k_0+\mathsf{N}}(\check{S},\check{E})\right)_i \right\}$ and $p = card(\mathscr{J})$.
If $p = \mathsf{N}$, then $(\widetilde{S},\widetilde{E}) \in \mathscr{X}$ defined by:

1. $\widetilde{E} = \check{E}$,
2. $\forall k \leqslant k_0, \widetilde{S}^k = \check{S}^k$,
3. $\forall k \in [\![1,\mathsf{N}]\!], \widetilde{S}^{k_0+k} = k$,
4. $\forall k > k_0 + \mathsf{N}, \widetilde{S}^k = 1$.

satisfies $d((\widetilde{S},\widetilde{E});(\check{S},\check{E})) < \delta$ and $\forall i \in [\![1,\mathsf{N}]\!], \mathscr{E}\left(G_{f_0}^{k_0+\mathsf{N}}(\widetilde{S};\widetilde{E})\right)_i \neq \mathscr{E}\left(G_{f_0}^{k_0+\mathsf{N}}(\check{S};\check{E})\right)_i$, so the result is obtained.
Else, let $j_1 < j_2 < ... < j_p$ be the elements of $\mathscr{J}$ and $j_0 \notin \mathscr{J}$. Then $\widetilde{X} = (\widetilde{E},\widetilde{S}) \in \mathscr{X}$ defined by

1. $\widetilde{E} = \check{E}$,
2. $\forall k \leqslant k_0, \widetilde{S}^k = \check{S}^k$,
3. $\forall k \in [\![1,p]\!], \widetilde{S}^{k_0+k} = j_k$,
4. $\forall k \in \mathbb{N}^*, \widetilde{S}^{k_0+p+k} = j_0$.

is such that $d(\check{X},\widetilde{X}) < \delta$. In addition, $\forall i \in [\![1,p]\!], \mathscr{E}\left(G_{f_0}^{k_0+\mathsf{N}}(\check{X})\right)_{j_i} \neq \mathscr{E}\left(G_{f_0}^{k_0+\mathsf{N}}(\widetilde{X})\right)_{j_i}$, because:

- $\forall i \in [\![1,\mathsf{N}]\!], \mathscr{E}\left(G_{f_0}^{k_0}(\check{X})\right)_i = \mathscr{E}\left(G_{f_0}^{k_0}(\widetilde{X})\right)_i$, due to the definition of k_0.
- $\forall i \in [\![1,p]\!], j_i \in \mathscr{J} \Rightarrow \mathscr{E}\left(G_{f_0}^{k_0+\mathsf{N}}(\check{X})\right)_{j_i} = \mathscr{E}\left(G_{f_0}^{k_0}(\check{X})\right)_{j_i}$, according to the definition of $\mathscr{J}$.
- $\forall i \in [\![1,p]\!], j_i$ appears exactly one time in $\widetilde{S}^{k_0}, \widetilde{S}^{k_0+1}, ..., \widetilde{S}^{k_0+\mathsf{N}}$, so

$$\mathscr{E}\left(G_{f_0}^{k_0+\mathsf{N}}(\widetilde{X})\right)_{j_i} \neq \mathscr{E}\left(G_{f_0}^{k_0}(\widetilde{X})\right)_{j_i}.$$

Lastly, $\forall i \in [\![1,\mathsf{N}]\!] \setminus \{j_0, j_1, ..., j_p\}, \mathscr{E}\left(G_{f_0}^{k_0+\mathsf{N}}(\widetilde{X})\right)_i \neq \mathscr{E}\left(G_{f_0}^{k_0+\mathsf{N}}(\check{X})\right)_i$, because:

- $\forall i \in [\![1,\mathsf{N}]\!], \mathscr{E}\left(G_{f_0}^{k_0}(\check{X})\right)_i = \mathscr{E}\left(G_{f_0}^{k_0}(\widetilde{X})\right)_i$,
- $i \notin \mathscr{J} \Rightarrow \mathscr{E}\left(G_{f_0}^{k_0+\mathsf{N}}(\check{X})\right)_i \neq \mathscr{E}\left(G_{f_0}^{k_0}(\check{X})\right)_i$,
- $i \notin \{\widetilde{S}^{k_0}, \widetilde{S}^{k_0+1}, ..., \widetilde{S}^{k_0+\mathsf{N}}\} \Rightarrow \mathscr{E}\left(G_{f_0}^{k_0+\mathsf{N}}(\widetilde{X})\right)_i = \mathscr{E}\left(G_{f_0}^{k_0}(\widetilde{X})\right)_i$.

So, in this case, $\forall i \in [\![1,\mathsf{N}]\!] \setminus \{j_0\}, \mathscr{E}\left(G_{f_0}^{k_0+\mathsf{N}}(\widetilde{S};\widetilde{E})\right)_i \neq \mathscr{E}\left(G_{f_0}^{k_0+\mathsf{N}}(\check{S};\check{E})\right)_i$ and the result of sensitivity is still obtained.

4.2.3 Expansivity

In this section we offer the proof that chaotic iterations are expansive [16] when f_0 is the update function:

Theorem 8. *$(\mathscr{X}, G_{f_0})$ is an expansive chaotic system. Its constant of expansivity is equal to 1.*

Proof. If $(S,E) \neq (\check{S};\check{E})$, then:

- Either $E \neq \check{E}$ and so at least one cell is not in the same state in E and $\check{E}$. Consequently the distance between (S,E) and $(\check{S};\check{E})$ is greater or equal to 1.
- Or $E = \check{E}$. So the strategies S and $\check{S}$ are not equal. Let n_0 be the first index such that the terms S and $\check{S}$ differ. Then $\forall k < n_0, G_{f_0}^{n_0}(S,E) = G_{f_0}^{k}(\check{S},\check{E})$, and $G_{f_0}^{n_0}(S,E) \neq G_{f_0}^{n_0}(\check{S},\check{E})$.
  As $E = \check{E}$, the cell that has changed in E at the n_0-th iterate is not the same than the cell that has changed in $\check{E}$, so the distance between $G_{f_0}^{n_0}(S,E)$ and $G_{f_0}^{n_0}(\check{S},\check{E})$ is greater or equal to 2.

So the expansivity property is established.

Remark 2. Expansivity is a kind of avalanche effect: any initial error is always magnified when iterating the system.

Remark 3. $(\mathscr{X}, G_{f_0})$ is not A-expansive, for any $A > 1$: let us consider two points $X = (E,S)$ and $X' = (E',S')$ with the same strategy ($S = S'$) and only one different cell ($d_e(E,E') = 1$). So, $\forall n \in \mathbb{N}, d_e\left(\mathscr{E}\left(G_{f_0}^n(X)\right), \mathscr{E}\left(G_{f_0}^n(X')\right)\right) = 1$.

4.3 Topological Entropy

Another important tool to measure the chaotic behavior of a dynamical system is the topological entropy, which is defined for compact topological spaces. Before studying the entropy of CIs, we must then check that $(\mathscr{X}, d)$ is compact.

4.3.1 Compacity Study

In this section, we will prove that $(\mathscr{X}, d)$ is a compact topological space, in order to study its topological entropy later. Firstly, as $(\mathscr{X}, d)$ is a metric space, it is separated. It is however possible to give a direct proof of this result:

Theorem 9. *$(\mathscr{X}, d)$ is a separated space.*

Proof. Let $(E,S) \neq (\hat{\text{E}}, \hat{\text{S}})$ two points of $\mathscr{X}$.

1. If $E \neq \hat{\text{E}}$, then the intersection between the two balls $\mathscr{B}\left((E,S), \frac{1}{2}\right)$ and $\mathscr{B}\left((\hat{\text{E}}, \hat{\text{S}}), \frac{1}{2}\right)$ in empty.
2. Else, it exists $k \in \mathbb{N}$ such that $S^k \neq \hat{\text{S}}^k$, then the balls $\mathscr{B}\left((E,S), 10^{-(k+1)}\right)$ and $\mathscr{B}\left((\hat{\text{E}}, \hat{\text{S}}), 10^{-(k+1)}\right)$ can be chosen.

The sequential characterization of the compacity for metric spaces can now be used to obtain the following result.

Theorem 10. *$(\mathscr{X}, d)$ is a compact metric space.*

Proof. Let $(E^n, S^n)_{n\in\mathbb{N}}$ be a sequence of $\mathscr{X}$.

1. A state $E^{\tilde{n}}$ which appears an infinite number of time in this sequence can be found. Let
$$I = \{(E^n, S^n)/E^n = E^{\tilde{n}}\}.$$
For all $(E,S) \in I$, $S_0^n \in [\![1,\mathsf{N}]\!]$, and I is an infinite set. Then $\tilde{\mathrm{k}} \in [\![1,\mathsf{N}]\!]$ can be found such that an infinite number of strategies of I starts with $\tilde{\mathrm{k}}$.
Let n_0 be the smallest integer such that $E^n = E^{\tilde{n}}$ and $S_0^n = \tilde{\mathrm{k}}$.
2. The set
$$I' = \{(E^n, S^n)/E^n = E^{n_0} \text{ and } S_0^n = S_0^{n_0}\}$$
is infinite, then one of the elements of $[\![1,\mathsf{N}]\!]$ will appear an infinite number of times in the S_1^n of I': let us call it $\tilde{1}$.
Let n_1 be the smallest n such that $(E^n, S^n) \in I'$ and $S_1^n = \tilde{1}$.
3. The set
$$I'' = \{(E^n, S^n) | E^n = E^{n_0} \text{ and } S_0^n = S_0^{n_0} \text{ and } S_1^n = S_1^{n_1}\}$$
is infinite, *etc.*

Let $l = \left(E^{n_0}, \left(S_k^{n_k}\right)_{k\in\mathbb{N}}\right)$, then the subsequence (E^{n_k}, S^{n_k}) converges to l.

4.3.2 Topological Entropy

Let (X,d) be a compact metric space and $f : X \to X$ be a continuous map. For each natural number n, a new metric d_n is defined on X by

$$d_n(x,y) = \max\{d(f^i(x), f^i(y)) : 0 \leq i < n\}.$$

Given any $\varepsilon > 0$ and $n \geqslant 1$, two points of X are ε-close with respect to this metric if their first n iterates are ε-close.

This metric allows one to distinguish in a neighborhood of an orbit the points that move away from each other during the iteration from the points that travel together. A subset E of X is said to be (n,ε)-separated if each pair of distinct points of E is at least ε apart in the metric d_n. Denote by $H(n,\varepsilon)$ the maximum cardinality of an (n,ε)-separated set,

Definition 9. The *topological entropy* of the map f is defined by (see e.g. [1] or [8])

$$h(f) = \lim_{\varepsilon\to 0}\left(\limsup_{n\to\infty}\frac{1}{n}\log H(n,\varepsilon)\right).$$

We have the result,

Theorem 11. *Entropy of $(\mathscr{X}, G_f)$ is infinite.*

Proof. Let $\mathrm{E}, \check{\mathrm{E}} \in \mathbb{B}^{\mathsf{N}}$ such that $\exists i_0 \in [\![1,N]\!], \mathrm{E}_{i_0} \neq \check{\mathrm{E}}_{i_0}$. Then, $\forall \mathrm{S}, \check{\mathrm{S}} \in \mathscr{S}$,

$$d((\mathrm{E},\mathrm{S});(\check{\mathrm{E}},\check{\mathrm{S}})) \geqslant 1$$

But the cardinal c of $\mathscr{S}$ is infinite, then $\forall n \in \mathbb{N}, c > e^{n^2}$.

Then for all $n \in \mathbb{N}$, the maximal number $H(n,1)$ of $(n,1)-$separated points is greater than or equal to e^{n^2}, so

$$h_{top}(G_f,1) = \overline{lim}\frac{1}{n}log\left(H(n,1)\right) > \overline{lim}\frac{1}{n}log\left(e^{n^2}\right) = \overline{lim}\ (n) = +\infty.$$

But $h_{top}(G_f,\varepsilon)$ is an increasing function when ε is decreasing, then

$$h_{top}\left(G_f\right) = \lim_{\varepsilon \to 0} h_{top}(G_f,\varepsilon) > h_{top}(G_f,1) = +\infty,$$

which concludes the evaluation of the topological entropy of G_f.

We have proven that it is possible to find f, such that chaotic iterations generated by f can be described by a chaotic and entropic map on a topological space in the sense of Devaney. We have considered a finite set of states $\mathbb{B}^{\mathsf{N}}$ and a set $\mathbb{S}$ of strategies composed by an infinite number of infinite sequences. In the following section we will discuss the impact of these assumptions in the context of the finite set of machine numbers.

5 Chaos in a Finite State Machine

Let us now explain how it is possible to have a true chaos in a finite state machine.

5.1 A Program with a Chaotic Behavior

In the Section 3 we have proven that discrete chaotic iterations can be put in the field of discrete dynamical systems:

$$\begin{cases} x^0 \in \mathscr{X} \\ x^{n+1} = G_f(x^n), \end{cases}$$

where $(\mathscr{X},d)$ is a metric space and G_f is a continuous function. Thus, it becomes possible to study the topological behavior of those chaotic iterations. Precisely, it has been proven that if the iterate function is based on the vectorial logical negation f_0, then chaotic iterations generate chaos according to Devaney. Therefore chaotic iterations, as Devaney's topological chaos, satisfy: sensitive dependence on the initial conditions, unpredictability, indecomposability, and uniform repartition. Additionally, G_{f_0} has been proven to be expansive and topologically mixing, and its topological entropy has been computed. Our intention is now to use these chaotic iterations that are highly unpredictable, to build programs in the computer science security field. Furthermore, we will give in Section 7 a link between CIs and artificial neural networks, thus it is possible to make them behave chaotically.

Up to now, most of computer programs presented as chaotic lose their chaotic properties while computing in the finite set of machine numbers. The algorithms that have been presented as chaotic usually act as follows. After having received its initial state, the machine works alone with no interaction with the outside world. Its outputs only depend on the different states of the machine. The main problem which prevents speaking about chaos in this particular situation is that when a finite state machine reaches a same internal state twice, the two future evolutions are identical. Such a machine always finishes by entering into a cycle while iterating. This highly predictable behavior cannot be set as chaotic, at least as expressed by Devaney. Some attempts to define a discrete notion of chaos have been proposed, but they are not completely satisfactory and are less recognized than the notions exposed in this paper.

The stated problem can be solved in the following way. The computer must generate an output O computed from its current state E *and* the current value of an input S, which changes at each iteration (Fig. 1). Therefore, it is possible that the machine presents the same state twice, but with two future evolutions completely different, depending on the values of the input. By doing so, we thus obtain a machine with a finite number of states, which can evolve in infinitely different ways, due to the new values provided by the input at each iteration. Thus such a machine can behave chaotically, as defined in the Devaney's formulation.

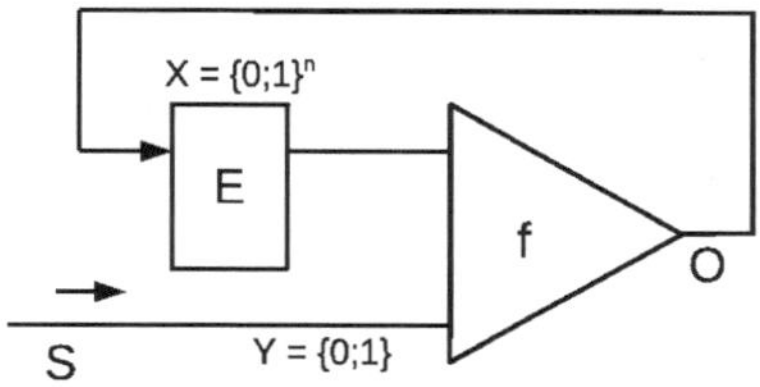

Fig. 1 A chaotic finite-state machine. At each iteration, a new value is taken from the outside world (S). It is used by f as input together with the current state (E).

5.2 The Practical Case of Finite Strategies

It is worthwhile to notice that even if the set of machine numbers is finite, we deal in practice with the *infinite* set of strategies that have *finite but unbounded lengths*. Indeed, it is not necessary to store all of the terms of these strategies in the memory. Only its nth term (an integer less than or equal to N) has to be stored at the nth step, as it is illustrated in the following example. Let us suppose that a given text is input from the outside world into the computer character by character and that the current term of the strategy is computed from the ASCII code of the current stored character. Since the set of all possible texts of the outside world is infinite and the number of their characters is unbounded, we work with an infinite set of finite but unbounded strategies.

In the computer science framework, we also have to deal with a finite set of states of the form $\mathbb{B}^{\mathsf{N}}$ and as stated before an infinite set $\mathbb{S}$ of strategies. The sole difference with the theoretical study is that instead of being infinite the sequences of S are finite with unbounded length, as any reasonable program must obviously finish one day.

The proofs of continuity and transitivity stated previously are independent of the finiteness of the length of strategies (sequences of $\mathbb{S}$). Sensitivity can be proven too in this situation (see Section 4.2.2). So even in the case of finite machine numbers, we have the two fundamental properties of chaos: sensitivity and transitivity, which respectively implies unpredictability and indecomposability (see [13], p.50). The regularity supposes that the sequences are of infinite lengths. To obtain the analogous of regularity in the context of finite sets, we can for example define a notion of *periodic but finite* sequences.

Definition 10. A strategy $S \in \mathbb{S}$ is said to be *periodic but finite* if S is a finite sequence of length n and if there exists a divisor p of n, $p \neq n$, such that $\forall i \leqslant n-p, S^i = S^{i+p}$. A point $(E,S) \in \mathscr{X}$ is said to be *periodic but finite*, if its strategy S is periodic but finite.

In this situation, $(1,2,1,2,1,2,1,2)$ (p=2) and $(2,2,2)$ (p=1), are periodic but finite. This definition can be interpreted as the analogous of periodicity definition on finite strategies. Following the proof of regularity (Section 3.2.1), it can be proven that the set of periodic but finite points is dense on $\mathscr{X}$, hence obtaining a desired element of regularity in finite sets, as quoted by Devaney ([13], p.50): “two points arbitrary close to each other could have completely different behaviors, the one could have a cyclic behavior as long as the system iterates while the trajectory of the second could ‘visit’ the whole phase space”. It should be recalled that the regularity was introduced by Devaney in order to counteract the effects of transitivity: two points close to each other can have fundamentally different behaviors.

In the following we explain how to use chaotic iterations in the computer science security field, by using an illustrative example. In this sense, we show two different ways to compute chaotic hash functions, the second one using neural networks.

6 Hash Functions with Topological Chaos Properties

In this section, a concrete example of a chaotic program is given in the computer science security field.

6.1 Introduction

The use of chaotic maps to generate hash algorithms has seen several developments in recent years. In [14] for example, a digital signature algorithm based on an elliptic curve and chaotic mapping is proposed to strengthen the security of an elliptic curve digital signature algorithm. Other examples of the generation of a hash function using chaotic maps can be found in, *e.g.*, [24, 26, 19]. Neural networks that have learned a continuous chaotic map have been proposed too in recent years [17], to achieve hash functions requirements.

Note that using any chaotic map does not guarantee that the resulting hash function would behave chaotically too. To the best of our knowledge, this point is not discussed in these referenced papers, however it should be considered as important. We define in this section a new way to construct hash functions based on chaotic iterations. As a consequence of the theory presented before, the generated hash functions satisfy various topological chaos properties. Thus, properties required for hash functions are guaranteed by our approach. For example, the avalanche criterion is deduced from the expansivity property.

6.2 A Chaotic Hash Function

In this section, we explain a new way to obtain a digest of a digital medium described by a binary sequence. It is based on chaotic iterations and satisfies various topological chaos properties. The hash value will be the last state of some chaotic iterations: the initial state X_0, finite strategy S, and iterate function must then be defined.

The initial condition $X_0 = (S, E)$ is composed by a $\mathsf{N} = 256$ bits sequence E and a chaotic strategy S. In the following section, we describe in detail how to obtain this initial condition from the original medium.

6.2.1 How to obtain E

The first step of our algorithm is to transform the message in a normalized 256 bits sequence E. To illustrate this step inspired by SHA-1, we state that our original text is: "*The original text*". Each character of this string is replaced by its ASCII code (on 7 bits). Then, we add a 1 at the end of this string.

```
10101001 10100011 00101010 00001101 11111100 10110100
11100111 11010011 10111011 00001110 11000100 00011101
00110010 11111000 11101001
```

So, the binary value (1111000) of the length of this string (120) is added, with another 1:

```
10101001 10100011 00101010 00001101 11111100 10110100
11100111 11010011 10111011 00001110 11000100 00011101
00110010 11111000 11101001 11110001
```

This string is inverted (the last bit is now the first one) and the two new substrings are concatenated. This gives:

```
10101001 10100011 00101010 00001101 11111100 10110100
11100111 11010011 10111011 00001110 11000100 00011101
00110010 11111000 11101001 11110001 00011111 00101110
00111110 10011001 01110000 01000110 11100001 10111011
10010111 11001110 01011010 01111111 01100000 10101001
10001011 0010101
```

So, we obtain a multiple of 512, by duplicating this string enough and truncating at the next multiple of 512. This string in which the whole original text is contained, is denoted by D.

Finally, we split the new string into blocks of 256 bits and apply the exclusive-or function, obtaining a 256 bits sequence in a manner inspired by the SHA-X algorithms.

```
11111010 11100101 01111110 00010110 00000101 11011101
00101000 01110100 11001101 00010011 01001100 00100111
01010111 00001001 00111010 00010011 00100001 01110010
01000011 10101011 10010000 11001011 00100010 11001100
10111000 01010010 11101110 10000001 10100001 11111010
10011101 01111101
```

In the context of Subsection 6.2, $\mathsf{N} = 256$, and E is the above obtained sequence of 256 bits: the given message has been compressed into a 256 binary string.

We now have the definitive length of our digest. Note that a lot of texts have the same normalized string. This is not a problem because the strategy we will build depends on the whole text too, in such a way that two different texts lead to two different strategies. Let us now build the strategy S.

6.2.2 How to Choose S

To obtain the strategy S, an intermediate sequence (u^n) is constructed from D as follows:

- D is split into blocks of 8 bits. Then u^n is the decimal value of the nth block.
- A circular rotation of one bit to the left is applied to D (the first bit of D is put on the end of D). Then the new string is split into blocks of 8 bits another time. The decimal values of those blocks are added to (u^n).
- This operation is repeated again 6 times.

It is now possible to build the strategy S:

$$S^0 = u^0, \quad S^n = (u^n + 2 \times S^{n-1} + n)\ (mod\ 256).$$

S will be highly dependent to the changes of the original text, because $\theta \longmapsto 2\theta\ (mod\ 1)$ is known to be chaotic as defined by Devaney's theory [13].

6.2.3 How to Construct the Digest

To construct the digest, chaotic iterations are done with initial state X^0,

$$f: \begin{array}{ccc} [\![1,256]\!] & \longrightarrow & [\![1,256]\!] \\ (E_1,\dots,E_{256}) & \longmapsto & (\overline{E_1},\dots,\overline{E_{256}}), \end{array}$$

as iterate function, and S for the chaotic strategy.

The result of those iterations is a 256 bits vector. Its components are taken 4 bits at a time and translated into hexadecimal numbers, to obtain the hash value:

```
63A88CB6AF0B18E3BE828F9BDA4596A6A13DFE38440AB9557DA1C0C6B1EDBDBD
```

To compare, if instead of using the text "*The original text*" we took "*the original text*", the hash function returns:

```
33E0DFB5BB1D88C924D2AF80B14FF5A7B1A3DEF9D0E831194BD814C8A3B948B3
```

In this paper, the generation of hash value is done with the vectorial Boolean negation f_0 defined in eq. (2). Nevertheless, the procedure remains general and can be applied with any function f such that G_f is chaotic. In the following subsection, a complete example of the procedure is given.

6.3 Application Example

Consider two black and white images of size 64×64 in Fig. 2, in which the pixel in position (40,40) has been changed. In this case, the hash function returns:

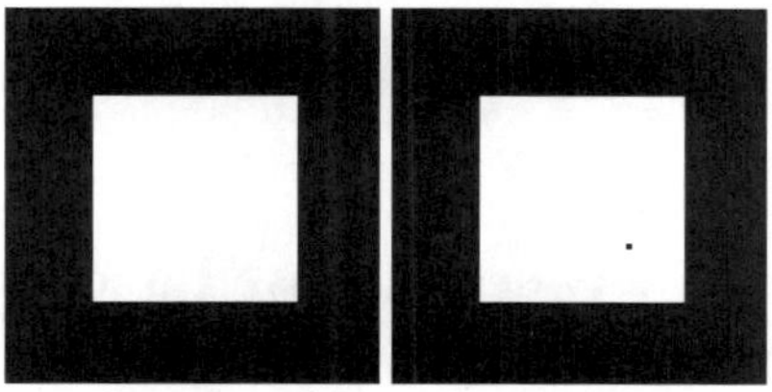

(a) Original image. (b) Modified image.

Fig. 2 Hash of some black and white images.

```
34A5C1B3DFFCC8902F7B248C3ABEFE2C9C9538E5104D117B399C999F74CF1CAD
```

for the Fig. 2(a) and

`5E67725CAA6B7B7434BE57F5F30F2D3D57056FA960B69052453CBC62D9267896`

for the Fig. 2(b).

Consider two 256 graylevel images of Lena (256×256 pixels) in figure 3, in which the grayscale level of the pixel in position (50,50) has been transformed from 93 (fig. 3(a)) to 94 (fig. 3(b)). In this case, the hash function returns:

(a) Original lena.

(b) Modified lena.

Fig. 3 Hash of some grayscale level images.

`FA9F51EFA97808CE6BFF5F9F662DCD738C25101FE9F7F427CD4E2B8D40331B89`

for the left Lena and

`BABF2CE1455CA28F7BA20F52DFBD24B76042DC572FCCA4351D264ACF4C2E108B`

for the right Lena.

These examples give an illustration of the avalanche effect obtained by this algorithm. A more complete study of the properties possessed by the hash functions and resistance under collisions will be studied in a future work.

7 A Chaotic Neural Network as Hash Function

A hash function can be achieved in two stages: the compression of the message (mapping a binary sequence of any length $n \in \mathbb{N}$ into a message of a fixed length belonging into $\mathbb{B}^{\mathsf{N}}$, for a given fixed length $\mathsf{N} \in \mathbb{N}$) and the hash of the compressed message [17]. As several compression functions have yet been proposed to achieve the first stage, we will only focus on the second stage and we will explain how to build a neural network that realize it. This neural network that hashes compressed messages will behave chaotically, as it is defined by the Devaney's theory.

Let us firstly explain how it is possible to build a neural network that behaves chaotically. Consider $f : \mathbb{B}^{\mathsf{N}} \longrightarrow \mathbb{N}^{\mathsf{N}}$ and a MLP which recognize F_f. That means,

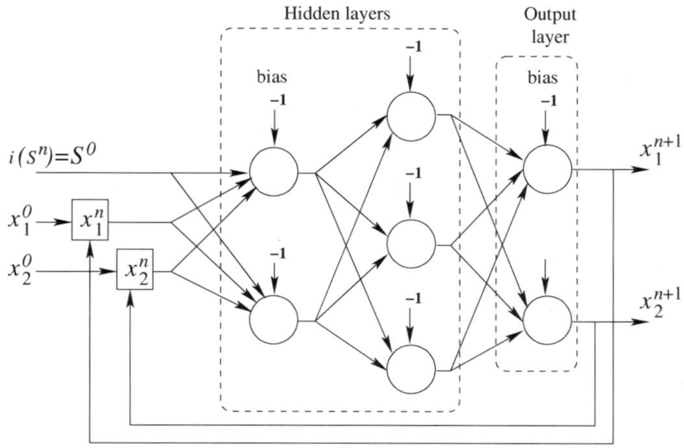

Fig. 4 Example of global recurrent neural network modeling function F_f such that $x^{n+1} = \left(x_1^{n+1}, x_2^{n+1}\right) = F_f\left(i(S^n), (x_1^n, x_2^n)\right)$

for all $(k,x) \in [\![1;\mathsf{N}]\!] \times \mathbb{B}^{\mathsf{N}}$, the response of the output layer to the input (k,x) is $F_f(k,x)$. We thus connect the output layer to the input one as it is depicted in Figure 4, leading to a global recurrent artificial neural network (ANN) working as follows [6]:

- At the initialization stage, the ANN receives a Boolean vector $x^0 \in \mathbb{B}^{\mathsf{N}}$ as input state, and $S^0 \in [\![1;\mathsf{N}]\!]$ in its input integer channel $i()$. Thus, $x^1 = F_f(S^0, x^0) \in \mathbb{B}^{\mathsf{N}}$ is computed by the neural network.
- This state x^1 is published as an output. Additionally, x^1 is sent back to the input layer, to act as Boolean state in the next iteration.
- At iteration number n, the recurrent neural network receives the state $x^n \in \mathbb{B}^{\mathsf{N}}$ from its output layer and $i(S^n) \in [\![1;\mathsf{N}]\!]$ from its input integer channel $i()$. It can thus calculate $x^{n+1} = F_f(i(S^n), x^n) \in \mathbb{B}^{\mathsf{N}}$, which will be the new output of the network.

Obviously, this particular MLP produce exactly the same values than CIs with update function f. That is, such MLPs are equivalent to CIs with f as update function. However, the compression stage of the hash function presented in the previous section can be resumed to making chaotic iterations over the compressed message. As chaotic iterations can be obtained with a neural network, we can thus realize this stage with a (chaotic) neural network. Finally, it is important to remark that the proposed hash function can be implemented into a global neural network, as various compression neural networks can be found in the literature [18, 21, 12]: we just have to replace our compression stage, inspired by SHA-X, with one of these compression ANN.

8 Conclusion

In this paper, a new approach to compute programs with a chaotic behavior is proposed. This approach is based on the well-known Devaney's topological chaos. The algorithms which are of iterative nature are based on the so-called chaotic iterations. This is achieved by establishing a link between the notions of topological chaos and chaotic iterations. Indeed, we are not interested in stable states of such iterations as it has always been the case in the literature, but in their unpredictable behavior. After a solid theoretical study, we consider the practical implementation of the proposed algorithms by evaluating the case of finite sets. We study the behavior of the induced computer programs proving that it is possible to design true chaotic computer programs.

An application is proposed in the area of information security: a new hash function is presented, the security in this case is guaranteed by the unpredictability of the behavior of the proposed algorithms. The algorithms derived from our approach satisfy important properties of topological chaos such as sensitivity to initial conditions, uniform repartition (as a result of the transitivity), unpredictability, expansivity, and topological mixing. Moreover, its topological entropy is infinite. The results expected in our study have been experimentally checked. The choices made in this first study are simple: compression function inspired by SHA-1, negation function for the iteration function, *etc*. The aim was not to find the best hash function, but to give simple illustrated examples to prove the feasibility in using the new kind of chaotic algorithms in computer science. Finally, we have shown how the mathematical framework of topological chaos offers interesting qualitative and qualitative tools to study the algorithms based on our approach.

In future work, we will investigate other choices of iteration functions and chaotic strategies. We will try to characterize transitive functions. Other properties induced by topological chaos will be explored and their interest in the information security framework will be deepened.

References

1. Adler, R.L., Konheim, A.G., McAndrew, M.H.: Topological entropy. Trans. Amer. Math. Soc. 114, 309–319 (1965)
2. Bahi, J., Guyeux, C.: A new chaos-based watermarking algorithm. In: SECRYPT 2010, Int. Conf. on Security and Cryptography, pp. 455–458. SciTePress, Athens (2010)
3. Bahi, J., Guyeux, C., Wang, Q.: A novel pseudo-random generator based on discrete chaotic iterations. In: INTERNET 2009, 1-st Int. Conf. on Evolving Internet, Cannes, France, pp. 71–76 (August 2009)
4. Bahi, J.M., Guyeux, C.: Hash functions using chaotic iterations. Journal of Algorithms & Computational Technology 4(2), 167–181 (2010)
5. Bahi, J.M., Guyeux, C.: Topological chaos and chaotic iterations, application to hash functions. In: WCCI 2010, IEEE World Congress on Computational Intelligence, pp. 1–7. Barcelona, Spain (2010); Best paper award
6. Bahi, J.M., Guyeux, C., Salomon, M.: Building a chaotic proved neural network. CoRR, abs/1101.4351 (2011)

7. Banks, J., Brooks, J., Cairns, G., Stacey, P.: On devaney's definition of chaos. Amer. Math. Monthly 99, 332–334 (1992)
8. Bowen, R.: Entropy for group endomorphisms and homogeneous spaces. Trans. Amer. Math. Soc. 153, 401–414 (1971)
9. Chazan, D., Miranker, W.: Chaotic relaxation. Linear Algebra and its Applications, 199–222 (1969)
10. Cong, J., Jiang, Y., Qu, Z., Zhang, Z.: A Wavelet Packets Watermarking Algorithm Based on Chaos Encryption. In: Gavrilova, M.L., Gervasi, O., Kumar, V., Tan, C.J.K., Taniar, D., Laganá, A., Mun, Y., Choo, H. (eds.) ICCSA 2006. LNCS, vol. 3980, pp. 921–928. Springer, Heidelberg (2006)
11. Zhu, C., Liao, X., Li, Z.: Chaos-based multipurpose image watermarking algorithm. Wuhan University Journal of Natural Sciences 11, 1675–1678 (2006), doi:10.1007/BF02831848
12. Cramer, C., Gelenbe, E., Bakircioglu, H.: Video compression with random neural networks. In: International Workshop on Neural Networks for Identification, Control, and Robotics, p. 476 (1996)
13. Devaney, R.L.: An Introduction to Chaotic Dynamical Systems, 2nd edn. Addison-Wesley, Redwood City (1989)
14. Peng, F., Qiu, S.-S., Long, M.: A secure digital signature algorithm based on elliptic curve and chaotic mappings. Circuits Systems Signal Processing 24(5), 585–597 (2005)
15. Guyeux, C.: Le désordre des itérations chaotiques et leur utilité en sécurité informatique. PhD thesis, Université de Franche-Comté (2010)
16. Guyeux, C., Friot, N., Bahi, J.: Chaotic iterations versus spread-spectrum: chaos and stego security. In: IIH-MSP 2010, 6-th Int. Conf. on Intelligent Information Hiding and Multimedia Signal Processing, Darmstadt, Germany, pp. 208–211 (October 2010)
17. Li, Y., Deng, S., Xiao, D.: A novel hash algorithm construction based on chaotic neural network. Neural Computing and Applications, 1–9 (2010)
18. Mahoney, M.V.: Fast text compression with neural networks. In: Proceedings of the Thirteenth International Florida Artificial Intelligence Research Society Conference, pp. 230–234. AAAI Press (2000)
19. Peng, F., Qiu, S.-S., Long, M.: One way hash function construction based on two-dimensional hyperchaotic mappings. Acta Phys. Sinici. 54, 98–104 (2005)
20. Robert, F.: Discrete Iterations: A Metric Study. Springer Series in Computational Mathematics, vol. 6 (1986)
21. Rudenko, O., Snytkin, M.: Image compression based on the neural network art. Cybernetics and Systems Analysis 44, 797–802 (2008)
22. Zhou, C.S., Chen, T.L.: Extracting information masked by chaos and contaminated with noise: Some considerations on the security of communication approaches using chaos. Physics Letters A 234(6), 429–435 (1997)
23. Wang, Q., Bahi, J., Guyeux, C., Fang, X.: Randomness quality of CI chaotic generators. application to internet security. In: INTERNET 2010. The 2nd Int. Conf. on Evolving Internet, pp. 125–130. IEEE Computer Society Press, Valencia (2010); Best Paper award
24. Wang, X.M., Zhang, J.S., Zhang, W.F.: One-way hash function construction based on the extended chaotic maps switch. Acta Phys. Sinici. 52(11), 2737–2742 (2003)
25. Wu, X., Guan, Z.-H.: A novel digital watermark algorithm based on chaotic maps. Physics Letters A 365(5-6), 403–406 (2007)
26. Xiao, D., Liao, X., Wang, Y.: Improving the security of a parallel keyed hash function based on chaotic maps. Physics Letters A 373(47), 4346–4353 (2009)

SOM-Based Techniques towards Hierarchical Visualisation of Network Forensics Traffic Data

E.J. Palomo, D. Elizondo, E. Domínguez, R.M. Luque, and Tim Watson

Abstract. Current research is improving the quality and efficiency of digital investigation methods due to the continuous proliferation of digital crimes. This includes the use of software tools that can help with digital investigations. A novel method for the analysis and visualisation of network forensics traffic data, based on growing hierarchical self-organising maps (GHSOM), is presented. Self-organising maps have shown to be successful for the analysis of highly-dimensional input data in data mining applications as well as for data visualisation. Moreover, the hierarchical architecture of the GHSOM is more flexible than a single SOM in the adaptation process to input data, capturing inherent hierarchical relationships among them. To evaluate the performance of this method in the field of network forensics, traffic data has been clustered and visualised in a hierarchical fashion to enhance the ability of digital forensics to find evidence of attacks or anomalous behaviour in the network. Experimental results show the utility of this approach.

Keywords: network forensics, data clustering, hierarchical self-organisation, data visualisation.

E.J. Palomo · E. Domínguez · R.M. Luque
Department of Computer Science
E.T.S.I. Informatica, University of Malaga
Campus Teatinos s/n, 29071 – Malaga, Spain
e-mail: {ejpalomo,enriqued,rmluque}@lcc.uma.es

D. Elizondo · Tim Watson
Department of Computer Technology
Forsec, De Monfort University
The Gateway, LE1 9BH – Leicester, United Kingdom
e-mail: {elizondo,tw}@dmu.ac.uk

D.A. Elizondo et al. (Eds.): Computational Intelligence for Privacy and Security, SCI 394, pp. 75–95.
springerlink.com

1 Introduction

The proliferation of digital crimes, boosted by the growth of the Internet, has highlighted the importance of the field of digital forensics. Digital forensics tries to discover evidence of an attack or an incident that has occurred by analysing large volumes of data. However, handling this data in an efficient manner is far from trivial [1]. Consequently, more and more research is being conducted into ways of improving the quality and efficiency of digital investigations. For this reason, forensics can be thought as a data mining issue. From this point of view, unsupervised learning methods are especially appropriated for understanding or extracting unknown information from different data sources. The purpose of unsupervised learning methods is to discover and cluster together similar groups from the input data. This data is usually represented as feature vectors in a highly-dimensional space. Data belonging to one group is more similar than data belonging to different groups as defined by a similarity measure. This is also referred as data clustering [2], being especially useful when information about the input data is unavailable or unknown.

The self-organising map (SOM) is an unsupervised learning method that has been widely used for knowledge discovery, data mining, detection of inherent structures in highly-dimensional data and mapping this data into a two-dimensional representation space [3]. This mapping retains relationships within the input data and preserves its topology. The main advantage of this method is the visual understanding of data structures. Different visualisation techniques have been frequently used to achieve a better understanding of the data [4]. The SOM has several weaknesses. One is that the number and arrangement of the neurons has to be defined in advance; another is their inability to represent hierarchical relationships. The growing hierarchical SOM (GHSOM) was proposed by Rauber et al. [5] to solve both limitations. This neural network model has a hierarchical architecture divided into layers, where each layer is composed of different single SOMs with an adaptative architecture that is determined during the unsupervised learning process based on the input data.

The authors use the GHSOM model for the clustering and the visualisation of network traffic data achieving a hierarchical clustering of traffic data. The understanding of this hierarchical clustering is improved by applying different visualisation techniques to the maps at each layer. Thus, all this information can be interpreted by a forensic investigator in order to find digital evidences that an attack or an incident has taken place.

The remainder of this chapter is organised as follows. In section 2, a description of the GHSOM model and the training algorithm is provided. Section 3, gives a brief overview about network forensics and data visualisation. In section 4, an experiment and its results consisting of the analysis and the visualisation of network forensic traffic data with the GHSOM are presented. Some conclusions are presented in section 5.

2 Hierarchical Self-Organising Model

2.1 Introduction to the Model

The Growing Hierarchical Self-Organising Map (GHSOM) [5] has been utilised as a tool to perform hierarchical cluster analysis over network traffic data. The GHSOM is an artificial neural network that consists of several growing SOMs [6] arranged in layers, where the entire architecture (layers, maps and neurons per map) is automatically determined during the unsupervised learning process. Initially, the GHSOM consists of a single SOM containing 2x2 neurons. Once this map is trained, it can grow by inserting rows or columns of neurons until a certain level of detail in the representation of the data mapped onto the SOM is reached. After growing, neurons are examined to decide whether they are expanded or not. This way, neurons that represent the most heterogeneous clusters are expanded into a new map in the next layer of the hierarchy in order to provide a more detailed representation. When the training process has finished, the final architecture of the GHSOM reflects the inherent hierarchical relationships among input data. This architecture also shows the clusters in which the data was mapped, with data belonging to one cluster being more similar than data belonging to different clusters. An example of the architecture of a GHSOM is shown in Figure (1).

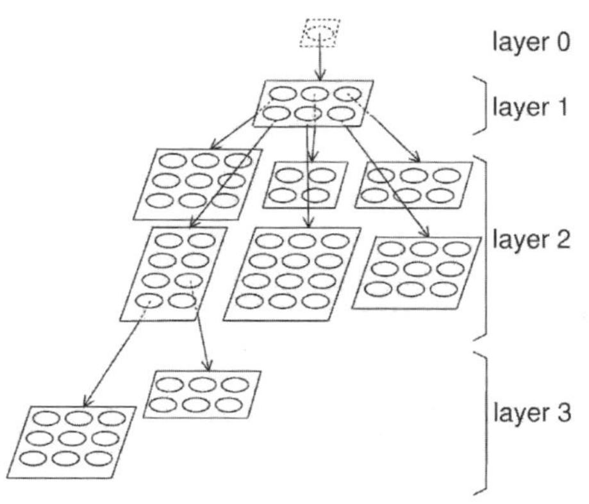

Fig. 1 The GHSOM architecture.

The adaptative growth process of a GHSOM is controlled by two parameters τ_1 and τ_2. These two parameters are used to control the growth of a map and the hierarchical growth of the GHSOM, respectively. This adaptation depends mainly on the quantisation error (qe) associated with each neuron. The qe of a neuron is a measure of the dissimilarity of the data mapped onto each neuron. The higher the value of qe, the higher the heterogeneity of the data cluster. Usually the qe is expressed in terms of the Euclidean distance between the input data and the weight vector of a neuron. However, in many real life problems not only quantitative features are present, but also qualitative features can be found. For instance, some qualitative features can be found when analysing network traffic data, such as the protocol type, whose values represent categories that do not have any order associated among them. It is common in this case to replace the qualitative features with $k-1$ dummy variables, where k corresponds to the number of possible values for the feature, which are binary variables that indicating whether the value is present or not. One drawback of this solution is the great amount of variables that must be inserted when the qualitative feature has a considerable number of values. This is even worse if the number of values is unknown, as is the case with the IP source address feature. Therefore, a distance for the qualitative features in order to compare two qualitative values was used. For the quantitative features the Euclidean distance was used. This new metric, that takes into account both type of features, is defined below.

Let $x_j = \binom{x_j^n}{x_j^c}$ be the jth input pattern, where x_j^n represents the quantitative (numerical) component of the feature vector and x_j^c the qualitative (categorical) component of the feature vector. The distance between two vectors v_1 and v_2, typically the weight (w) and input pattern (x) vectors, is defined in (1) as the 2-norm of the difference vector for the quantitative component and a binary vector for the qualitative component, as expressed in (2).

$$d(v_1, v_2) = ||\delta(v_1, v_2)|| \tag{1}$$

$$\delta(v_1, v_2) = \begin{pmatrix} v_1^n - v_2^n \\ v_1^c \oplus v_2^c \end{pmatrix} \tag{2}$$

where $\oplus$ is the logical operator, exclusive sum (xor), which generates a binary vector representing the distance between qualitative features. Therefore, for quantitative components the Euclidean distance is computed, whereas for qualitative components a distance for 0 or 1 is used whether the qualitative values are the same or not.

Based on the distance defined in (1), the qe of a neuron i is defined as follows:

$$qe_i = \sum_{x_j \in C_i} d(w_i, x_j) \tag{3}$$

where C_i is the set of input patterns mapped onto the neuron i, x_j is the jth input pattern from C_i, and w_i is the weight vector of the neuron i. Since the qe of a neuron is computed as the sum of the distances between the weight vector of a neuron and the input data mapped into the neuron (3), the larger the number of input data mapped, the larger the qe. Another alternative is to compute the mean quantisation error (mqe). Although the mqe may be more intuitive as a dissimilarity measure of a cluster, using the qe we can give more map space for the largest clusters, which is also referred as *magnification factor* [5]. Thus, the distribution of the data in populated neurons can be explored at a higher level of granularity, favouring understanding of the characteristics of the data.

Initially, the quantisation error at layer 0 must be computed as given in (4), where w_0 is the mean of the all input data I.

$$qe_0 = \sum_{x_j \in I} d(w_0, x_j) \tag{4}$$

The initial quantisation error (qe_0) measures the dissimilarity of all input data and it is utilised as the stopping criterion of the hierarchical growth (5), in which the quantisation error of a neuron i (qe_i) must be smaller than a fraction (τ_2) of the initial quantisation error (qe_0). Otherwise, the neuron is expanded into a new map at the next level of the hierarchy. Hence the τ_2 parameter controls the size of the hierarchy so the smaller this parameter is, the deeper the hierarchy will be.

$$qe_i < \tau_2 \cdot qe_0 \tag{5}$$

The newly created maps from expanded neurons need their weight vectors to be coherently initialised as proposed by Dittenbach et al. [7]. This initialisation provides a global orientation of the individual maps in the various layers of the hierarchy. Thus, the weight vectors of neurons mirror the orientation of the weight vectors of the neighbour neurons of their parents. The proposed initialisation computes the mean of the parent and its neighbours in their respective directions.

A new map, created from an expanded neuron, is trained as a single SOM. During the training of this map, the set of input patterns are those that were mapped onto the upper expanded neuron. In each iteration t, an input pattern is randomly selected from this data subset. The winning neuron of the map is the neuron with the smallest proposed distance in (1) to the input pattern, whose index r is defined in (6).

$$r(t) = \arg\min_i (d(w_i, x(t))) \tag{6}$$

The winner's weight vector is updated according to the expression (7) making a difference for the quantitative and qualitative components of the vector. For quantitative features, the amount of adaptation of the winning neuron is guided by a learning rate α, decreasing in time (7). In addition to the winner, the winner's neighbours are updated depending on a Gaussian neighbourhood function h_i and its distance to the winning neuron. This neighbourhood function reduces its neighbourhood kernel at each iteration. For qualitative features, only the winner is adapted with the weight vector that represents the mode of the set of input patterns mapped onto the winner C_i.

$$w_i(t+1) = \begin{pmatrix} w_i^n(t+1) \\ w_i^s(t+1) \end{pmatrix} = \begin{pmatrix} w_i^n(t) + \alpha(t)h_i(t)[x^n(t) - w_i^n(t)] \\ mode(C_i) \end{pmatrix} \quad (7)$$

Once the map is trained, the map can grow in which case, the training process starts over again. In order to know when to grow, the stopping criterion for the growth of a map (8) must be checked. If the mean quantisation error of a map m (MQE_m) is smaller than a certain fraction τ_1 of the quantisation error of the corresponding parent neuron u in the upper layer, the map stops growing. Otherwise, the map keeps growing to achieve a better level of representation of the data mapped onto the map. So the smaller the τ_1 parameter chosen is the larger the map is.

$$MQE_m < \tau_1 \cdot qe_u \quad (8)$$

The growth of a map is done by inserting a row or a column of neurons between two neurons, the neuron with the highest quantisation error e and its most dissimilar neighbour d. A row or a column is inserted if these two neurons are either in the same column or in the same row, respectively. The index of the neuron d is computed according to the expression (9), where Λ_e is the set of neurons neighbouring e.

$$d = \arg\max_i (d(w_e, w_i)), \qquad w_i \in \Lambda_e \quad (9)$$

2.2 Training Algorithm

The pseudo code of the training algorithm of the proposed GHSOM is defined as follows:

3 Network Forensics

At present, network security can be a critical problem. Cybercrime today is alive and well on the internet and growing both in scope and sophistication. Much progress is being made to protect both the internet and intranets using defensive mechanisms such as firewalls and intrusion detection systems, but the capacity of these solutions is limited and not sufficient to eliminate cyber attacks. The wide use of computer systems and networks brings a natural increase of criminal activity related to internet and computing devices, leading to a rapid raise of the field of digital forensics. Digital forensics is the use of scientific methods for the identification, preservation, extraction and documentation of digital evidence derived from digital sources to enable successful prosecution or proof of innocence [8]. Digital evidence, by definition, is information of probative value stored or transmitted in digital form [9]. Digital forensics can be divided into two main areas, computer forensics and network forensics. Computer forensics identifies evidence that particular

```
Input: X = x_1, x_2, ..., x_m
Output: W = w_1, w_2, ..., w_n

BEGIN

Initialise w_0 and then, qe_0
Create a map with 2x2 neurons
maps:= 1
For m:=1 To maps Do
    Repeat
        For l:=1 To λ Do
            Train the map during as a single SOM
        End
        Compute the qe of each neuron
        Compute the MQE_m
        If (MQE_m >= τ_1 · qe_u)
            Select the neurons e and d
            Insert a row or column of neurons between d and e
        End
    Until (MQE_m < τ_1 · qe_u)
    For i:=1 To neurons Do
        If (qe_i >= τ_2 · qe_0)
            Create a map with 2x2 neurons
            maps:= maps + 1
        End
    End
End

END
```

computers have been used in the perpetration of specific crimes [10], whereas network forensics serves to prevent attacks into systems or to find potential evidence after an attack or an incident has occurred [11]. Such attacks include probing, denial of service (DoS), user-to-root (U2R) and remote-to-local (R2L).

Network forensics is a research area that gathers evidence of legal or illegal acts by collecting and analysing the computer network traffic. Unlike other areas of digital forensics, network investigations deal with volatile and dynamic information. Nowadays, due to the increase of network traffic, not all the information captured or recorded will be useful for evidence or analysis. The network forensic process consist on capturing, recording and analysing network audit trails [11]. This data is often located in the logs of the devices dispersed throughout the network, for example, routers, switches, firewalls, Web servers, but can also be present on the traffic data itself. Examples of activities considered part of network forensics are the analysis of the logs of intrusion detection systems (IDSs) [12], the analysis of network traffic data [13] and the analysis of network devices themselves [14]. For the analysis of network traffic, the common practice is to archive all traffic and analyse subsets as necessary, which is also known as reconstructive traffic analysis [15].

Network forensics experts try to find the malicious users by collecting and analysing the intrusion or infringement evidence of computer crimes such as hacking. However, due to the increasing volume of network traffic in today's networks, it is infeasible to effectively store and query all the network information (packets) for extended periods of time, in order to allow analysis of evidence. Moreover, current digital forensic tools simply cannot handle large volumes of data in an efficient manner. In this sense, the forensic investigators have difficulty in analysing the state of the damaged systems. A simply digital investigation, analysing multiple computer systems, can easily consume an enormous amount of time. Since network forensics tasks involve the manual examination of this data, which can be a time-consuming and error-prone process [16], some tools have recently been developed in order to automate this methodology. Such tools include EnCase Forensic [17], Forensic Toolkit [18], SafeBack [19] and Pro Discover [20], amongst others. However, these tools have some limitations when large volumes of data need to be analysed [1]. For this reason, data mining techniques are excellent candidates to face this problem [21].

Particularly, the self-organising map (SOM) has the advantage of visualising multidimensional data in a meaningful way [4, 22]. According to Fei et al. [23], the SOM is proposed as a visualisation tool to support decision making by computer forensic investigators, whereas in a later work [24] the SOM is utilised to determine anomalous behaviours (or activities) among employees (or computer systems) in a far more efficient manner. The data mining and visualisation capabilities of the SOM, together with the hierarchy representation of the data provided by the GHSOM, are utilised here to help on network forensic investigations. Thus, low-dimensional views of highly-dimensional network forensics data can be visualised both to achieve a better understanding about the obtained data clusters and to help with the examination of these data to find potential evidence of computer crimes.

4 Experimental Results

In order to demonstrate the application of the GHSOM in the field of network forensics, an experiment and its results are described in this section. This experiment consists on the analysis and data visualisation of network traffic issued from the monitoring of network packets flowing over different subnets within an entire network infrastructure in a university environment, known that some attacks could have taken place. Network analysis and subsequent visualisation of resulting data is carried out by the GHSOM model. The resulting data visualisations are helpful with the forensic analysis process of the network.

4.1 Data Preparation

For the experiments, network packets were captured from two different subnets from a university network with the aid of the WireShark program. These packets correspond to five consecutive days of monitoring the network traffic. The resulting data set contains 149,601 samples, where each sample corresponds to a captured network packet. Seven input features were selected to represent a network packet, namely, IP source address, IP destination address, protocol type, date, time, packet length and delta time (time between captured packets). From these features, the first five are qualitative, whereas the last two are quantitative. This feature division must be taken into account when training the proposed GHSOM to perform a correct clustering, as explained in Section 2. For the date feature, only the weekday (a number between 1 and 7) was considered. Although the date feature is an ordinal variable having its values an associated order, it is also a qualitative variable since these values are discrete and, therefore, new values cannot be created. The packet length feature represents the number of bytes of a packet and the delta time feature is the time since the previous packet was captured.

4.2 Data Visualisation

Cluster analysis and visualisation of network traffic data was performed by training the GHSOM model with the network traffic data set. For this purpose, both the τ_1 and the τ_2 parameters were set to 0.1 since they achieved a good trade-off between the size of the maps and the depth of the hierarchy. This is important in order to get an understandable architecture that can be easily visualised and interpreted. After training, a neural network architecture that reflects the inherent hierarchical relationships among input data was obtained. In this neural network, neurons represent network traffic data clusters. Two different views of the resulting 3D architecture are shown in Figure 2. In the plot, the hierarchical topology of the clusters in which the input data was mapped can be seen. In fact, two layers were automatically generated for this purpose, producing three maps in the second layer. Each sphere represents a neuron. Numbers were assigned to each neuron. The expanded neurons are represented without horizontal lines. An arrow going out from the neuron indicates the map in which it was expanded. Note that neurons 1, 2 and 4 have been expanded into three maps in the second layer with 5x2, 3x4 and 2x5 neurons, respectively. These maps represent

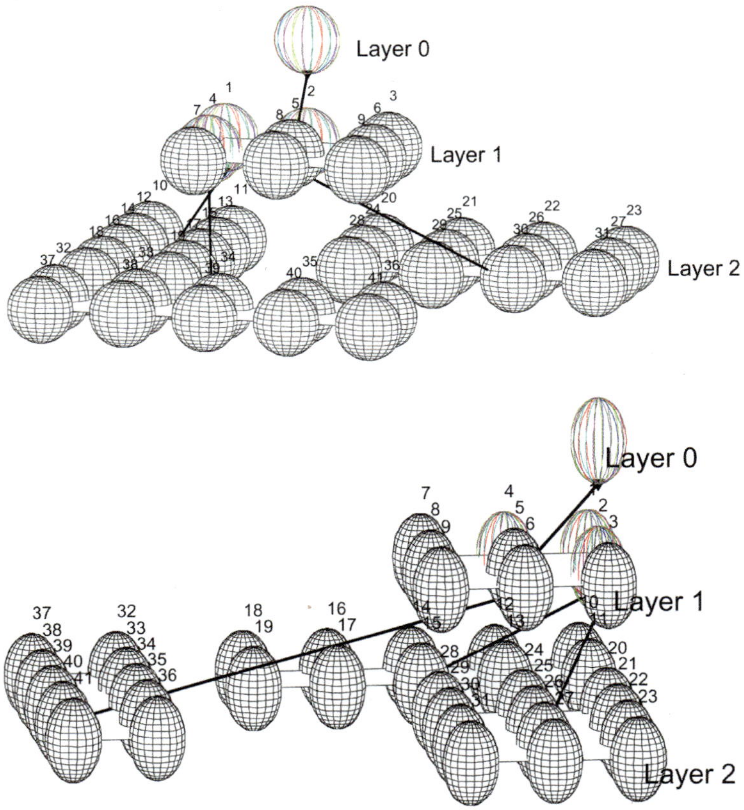

Fig. 2 3D architecture of the resulting GHSOM from two perspectives. Each sphere represents a neuron. Numbers were assigned to each neuron. Two layers were automatically generated, producing one map with 3x3 neurons in the first layer and three maps with 5x2, 3x4 and 2x5 neurons in the second layer that were expanded from neurons 1, 2 and 4, respectively. The expanded neurons are represented without horizontal lines. An arrow going out from the neuron indicates the map in which it was expanded.

subclusters from their respective parent clusters. Another possible visualisation of this architecture is shown in Figure 3, where the 3D architecture was projected in 2D. Each square represents a neuron, where the smaller the size of the square, the deeper the level of the hierarchy in which the neuron is (the biggest square represents the neuron at layer-0 map). By comparing the two representations of the neural architecture, a better understanding of the resulting topology, in which input data have been hierarchally self-organised, can be achieved.

As discussed above, the neural network architecture provides information about the topology of the clusters, i.e. how neurons from the same map or from different maps are connected with others. However, no information

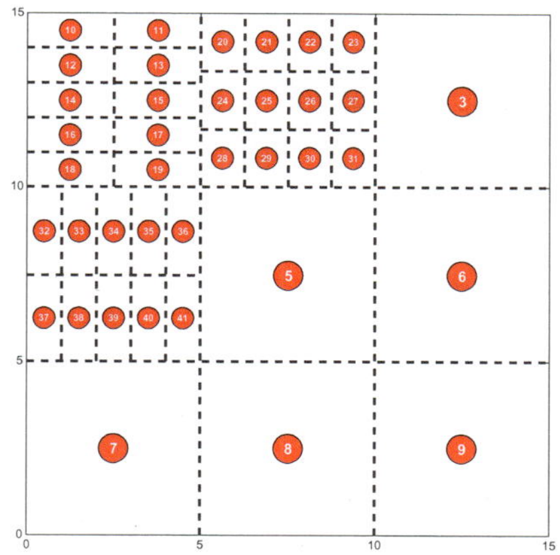

Fig. 3 Projection of the 3D architecture of the resulting GHSOM in 2D. Each square represents a neuron, where the smaller the size of the square, the deeper the level of the hierarchy in which the neuron is (the biggest square represents the neuron at layer-0 map). Numbers were assigned to each neuron. Two layers were automatically generated, producing one map with 3x3 neurons in the first layer and three maps with 5x2, 3x4 and 2x5 neurons in the second layer that were expanded from neurons 1, 2 and 4, respectively. The expanded neurons are those that were divided into more neurons.

about the input data is provided. Therefore, some visualisation techniques to show relationships between input data and neurons are considered. First of all, the plot of the input data hits for each map is presented. The number of samples mapped into a neuron indicates the size of the clusters and how these data are distributed amongst the different maps of the hierarchy. This mapping between samples and neurons is shown in Figures 4 and 5 for the first and second layer, respectively. A total of 149,601 samples were mapped into the layer-1 map. For the three layer-2 maps a total of 22,318, 50,868 and 30,408 samples were mapped, respectively. In the plots, each neuron shows the number of samples mapped and also the relative number of samples for each neuron by means of the size of a coloured patch. The neurons with a larger number of samples are the ones that have been expanded. This is due to the fact that the quantisation error was used instead of the mean quantisation error to capture finer differences in more densely populated clusters. This enables a better insight and understanding of the data set (see Section 2).

In Figure 5, some neurons are visible in two of the maps without any sample associated. These dead neurons give us an idea of the shape of the map and the distances between neurons, although that is not the purpose of the plot. In order to visualise the above, weight vectors together with input data are usually plotted. However, when the input space is highly-dimensional and qualitative features are present as occurred in our experiment, the U-matrix is used instead. The U-matrix (unified distance matrix) [25] visualises the

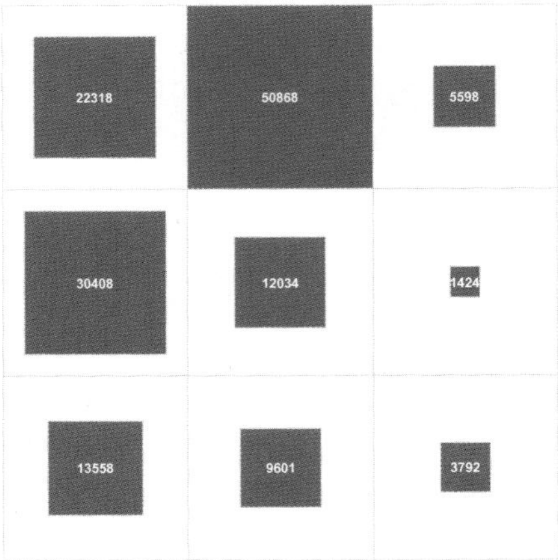

Fig. 4 Input data hits from layer-1 map. A total of 149,601 samples were mapped into the layer-1 map. Each neuron shows the number of samples mapped and also the relative number of samples for each neuron by means of the size of a coloured patch. The neurons with a larger number of samples are the ones that have been expanded.

distances between neighbouring neurons using different colours to represent distances. Darker colours represent larger distances, whereas lighter colours represent smaller distances. The U-matrix for the layer-1 map and those for the three layer-2 maps are depicted in Figures 6 and Figure 7, respectively. In these figures, the small hexagons represent the neurons, the lines connect neighbouring neurons and the greyscale in the regions containing the lines indicate the distance between neurons (the larger the distance, the darker the shade). In Figure 6, a lighter line surrounded by dark regions can be seen. This indicates that there is a group of neurons that are closer than the others, namely, neurons 1, 4, 7, 8, 5, 6 and 9 neurons. Also, it can be seen that neurons 2 and 3 are far away from the rest of the neurons, indicating that these clusters are more different than the others. The same analysis can be applied to the U-matrices for the three layer-2 maps shown in Figure 7. In the third map (the bottom one), the input data has been clustered into two main groups due to the darker line that split the map into two parts.

Another useful plot for the analysis and visualisation of network forensics traffic data is the component planes. The component planes are visualisations of the map weights of the neurons, showing the spread of values of each component (or feature) in all map neurons. Thus, possible correlations among features can be discovered by comparing one component plane with another. A component plane for each of the seven features present in the input data for the layer-1 map and the three maps in the second layer is shown in Figures 8 and 9, respectively. In these figures, larger weights are

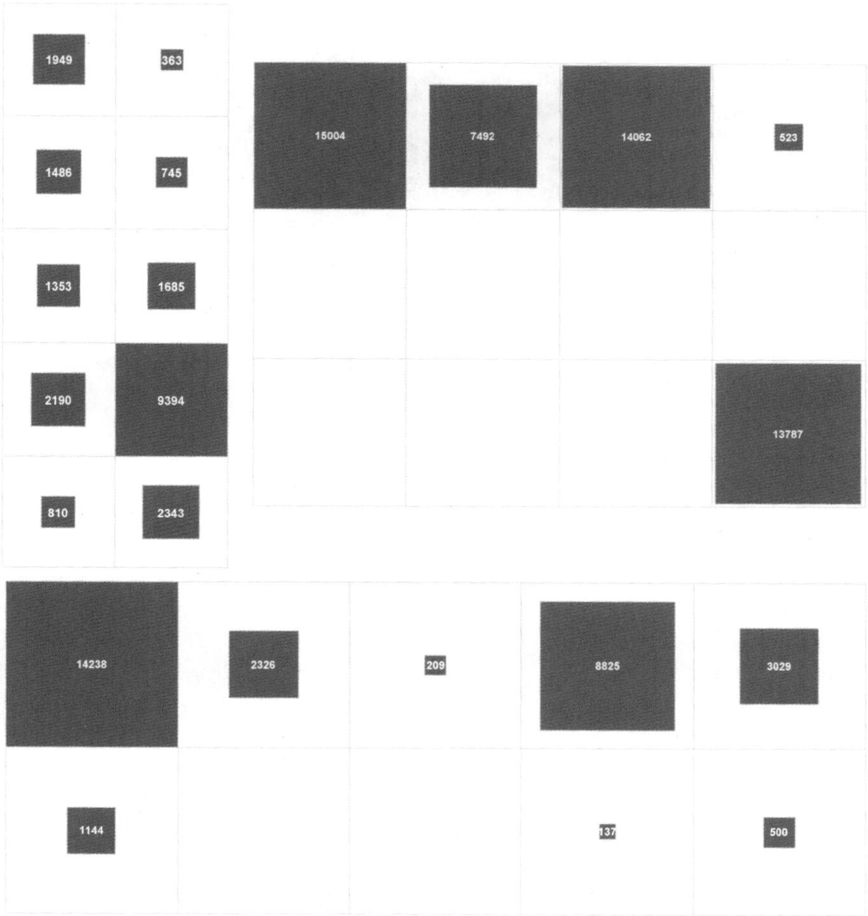

Fig. 5 Input data hits from the three layer-2 maps. A total of 22,318, 50,868 and 30,408 samples were mapped into the three layer-2 maps, respectively. Each neuron shows the number of samples mapped and also the relative number of samples for each neuron by means of the size of a coloured patch. Dead neurons (neurons without mapped samples) give us an idea of the shape of the map and the distances between neurons.

represented by darker shades. It should be remembered that, for quantitative components, the weight of a neuron represents the centroid of the input data mapped into that neuron, whereas for qualitative components it represents the mode of the same samples (see Equation 7). For qualitative components, the first five features of the network traffic data set, each neuron's weight value is associated to a different shade without any ordination. Therefore the greyscale only indicates that the values are different, except for the weekday and time features, in which case the greyscale represents an order. Thus, for

Fig. 6 U-matrix for layer-1 map. The small hexagons represent the neurons, the lines connect neighbouring neurons and the greyscale in the regions containing the lines indicates the distance between neurons (the larger the distance, the darker the shade).

weekday values the lightest shade corresponds to 'Monday' and the darkest to 'Sunday', whereas for time values, shades are assigned within the range [0-23] (the larger the value, the darker the shade).

Table 1 Post-processed component values for neurons of the layer-1 map.

Neuron	Origin	Protocol	Weekday	Time	Length (bytes)	Δ-time (secs)
1	Local	TCP	Friday	20-21	61.29	2.56
2	Local	UDP	Saturday	21-22	42.20	$4.99\text{x}10^{-4}$
3	France	TCP	Thursday	19-20	70.21	4.70
4	Local	TCP	Saturday	21-22	61.19	1.13
5	Local	ARP	Friday	20-21	50.01	5.19
6	Pakistan	UDP	Friday	20-21	43.08	$5.71\text{x}10^{-3}$
7	Local	ARP	Saturday	21-22	50.24	7.77
8	Local	ARP	Thursday	19-20	50.36	6.40
9	Pakistan	DEC DNA	Friday	20-21	68.38	1.72

By analysing the plot in Figure 8, it can be observed how the destination address, protocol and packet length features are highly correlated, since they have the same distribution in the map. This indicates that most of the packets with the same destination address have the same protocol associated and, consequently, the same packet length. This is not the same for the source address since the origin can be very diverse. The destination address is always an IP address of the university network. Note that packet length and

Fig. 7 U-matrixes for the three layer-2 maps. The small hexagons represent the neurons, the lines connect neighbouring neurons and the greyscale in the regions containing the lines indicates the distance between neurons (the larger the distance, the darker the shade).

delta time features are not correlated since packets with the same length can have different delta times. Special attention was paid to those clusters whose predominant (prototype) IP source address comes from the outside of the network. In fact, it was observed that most of the IP source addresses came from Pakistan in neurons 6 and 9 and from France in neuron 3. Likewise, the weekday and time features are correlated. For these components, neurons constitute three groups where the majority of the packets were captured on Thursday between 7 p.m. and 8 p.m. (neurons 3 and 8), Friday between 8 p.m. and 9 p.m. (neurons 1, 5, 6 and 9) and Saturday between 9 p.m. and 10 p.m. (neurons 2, 4 and 7) for each formed group. Taking into account the time lag of the packets that come from foreign countries, the weekday and time to consider in this analysis is Thursday between 8 p.m. and 9 p.m. for

Fig. 8 Component planes for layer-1 map. Larger weights are represented by darker shades. For qualitative components (source address, destination address, protocol, weekday and time), each neuron's weight value is associated to a different shade without any ordination. For quantitative components (packet length and delta time), shades has an ordination.

neuron 3 (Central European Time) and Saturday between 1 a.m. and 2 a.m. for neurons 6 and 9 (Pakistan Standard Time). Since most of the network activity belongs to the end of the week during the night, this information can be an evidence of anomalous behaviour for a forensic investigator. Especially for the source addresses that come from the outside of the network, which are present in neurons 3, 6 and 9. From these neurons, packets mapped into neuron 9 seem to be harmless since the predominant protocol is DEC DNA, a distance vector routing protocol in which routers periodically send routing updates to all neighbours by broadcasting their entire route tables, whereas the most numerous protocol in neuron 3 and 6 are TCP and UDP, respectively. Therefore, the most anomalous packets are those that were clustered in neurons 3 and 6, which were not expanded by the GHSOM model. This information extracted from the weight vectors of each neuron from the layer-1 map are summarised in Table 1. To facilitate the interpretation of the plots, dead neurons are removed.

Similarly, by observing the component planes of the three layer-2 maps shown in Figure 9, others suspicious packets that went unnoticed in the first layer can be identified. These maps were expanded from neurons 1, 2 and 4

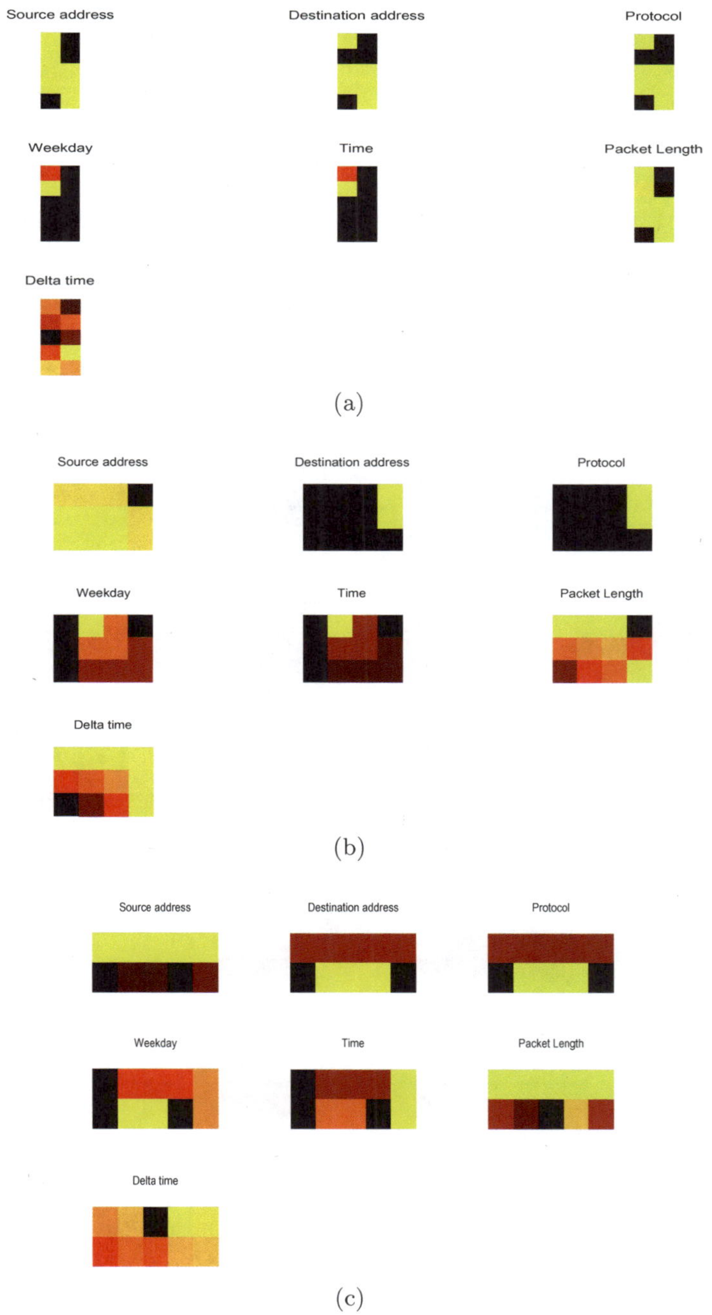

Fig. 9 Component planes for the three layer-2 maps: (a) map 1 (b) map 2 and (c) map 3 from parent neurons 1, 2 and 4 from the layer-1 map. Larger weights are represented by darker shades. For qualitative components (source address, destination address, protocol, weekday and time), each neuron's weight value is associated to a different shade without any ordination. For quantitative components (packet length and delta time), shades has an ordination.

Table 2 Post-processed component values for neurons of the layer-2 map 1.

Neuron	Origin	Protocol	Weekday	Time	Length (bytes)	Δ-time (secs)
1	Local	UDP	Thursday	19-20	60.01	2.78
2	Pakistan	TCP	Friday	20-21	79.01	7.36
3	Local	TCP	Wednesday	18-19	60.80	5
4	Pakistan	TCP	Friday	20-21	77.71	3.37
5	Local	UDP	Friday	20-21	60	8.71
6	Local	UDP	Friday	20-21	60	6.43
7	Local	UDP	Friday	20-21	60	4
8	Local	UDP	Friday	20-21	60	0.15
9	Pakistan	TCP	Friday	20-21	77.60	1.08
10	Local	UDP	Friday	20-21	60	2.07

Table 3 Post-processed component values for neurons of the layer-2 map 2.

Neuron	Origin	Protocol	Weekday	Time	Length (bytes)	Δ-time (secs)
1	Local	UDP	Saturday	21-22	42	$1.06\text{x}10^{-5}$
2	Local	UDP	Wednesday	15-16	42	$8\text{x}10^{-6}$
3	Local	UDP	Thursday	19-20	42	$1.44\text{x}10^{-5}$
4	Italy	DEC DNA	Saturday	21-22	63.36	0.07
12	Local	UDP	Friday	20-21	42	$1.23\text{x}10^{-5}$

Table 4 Post-processed component values for neurons of the layer-2 map 3.

Neuron	Origin	Protocol	Weekday	Time	Length (bytes)	Δ-time (secs)
1	Local	UDP	Saturday	21-22	60	1.69
2	Local	UDP	Thursday	19-20	60	1.04
3	Local	UDP	Thursday	19-20	60	6.40
4	Local	UDP	Thursday	19-20	60	0.03
5	Local	UDP	Wednesday	15-16	60	0.30
6	Italy	TCP	Saturday	21-22	80.27	3.41
9	Italy	DEC DNA	Saturday	21-22	64.19	1.13
10	Italy	TCP	Wednesday	15-16	79.90	0.91

(Figure 9(c)), which were heterogeneous enough to be analysed at a higher level of detail. From the first map (see Figure 9(a)), it can seen three clusters that are different to the rest, namely, neurons 2, 4 and 9, which represent packets whose IP source address came from Pakistan on Friday between 8 p.m. and 9 p.m. (corresponding to Saturday between 1 a.m. and 2 a.m. following the Pakistan Standard Time) using the protocol TCP. Note that these neurons are the less populated compared with the others (see Figure 5). The rest of the neurons are similar to the parent neuron. In the second map (see Figure 9(b)), the cluster that may contain anomalous packets is neuron 4.

This corresponds to the less dense cluster of the map with DEC DNA packets (from the mapped 50,868) from Italy on Saturday between 9 p.m. and 10 p.m. (between 10 p.m. and 11 p.m. following the Central European Time). Finally, the third map indicates three noteworthy possible anomalous clusters whose packets mainly came from Italy: neuron 6 with TCP packets on Saturday between 9 p.m. and 10 p.m. (between 10 p.m. and 11 p.m. in Italy), neuron 9 with DEC DNA packets on the same date and time than neuron 3, and neuron 10 with TCP packets on Wednesday between 3 p.m. and 4 p.m. (between 4 p.m. and 5 p.m. in Italy). Information gathered from the weight vectors of neurons from the three maps of the second layer, is shown in Tables 2 - 4.

5 Conclusions

A hierarchical visualisation of network forensic traffic data is provided using a GHSOM. The GHSOM is composed of several independent growing SOMs, arranged in a hierarchical architecture, that are automatically established during the unsupervised learning process. As a consequence, this hierarchical architecture provides a more flexible adaptation to input data than a single SOM and a better capture of the relationships among them. Therefore, the improvement in data modelling provides a better visualisation and understanding of the input data.

By using the GHSOM to cluster network traffic data, the provided hierarchy of SOMs can be analysed to extract relevant information about the mapped data. This may aid a network forensic investigator to better understand hidden relationships in the data sets. In our experiment, a data set obtained by capturing five days of traffic data from a university network environment was utilised to train the GHSOM. Different visualisation techniques were used to represent the different maps obtained. After considering all the data plots together with the hierarchical architecture, a better insight of the data was achieved and clusters of suspicious network behaviour could be identified along the hierarchy.

Acknowledgements. This work is partially supported by the Ministry of Science and Innovation of Spain under grant TIN2010-15351, project name Probabilistic self organizing models for the restoration of lossy compressed images and video.

References

1. Roussev III, V., Richard, G.G.: Breaking the performance wall: The case for distributed digital forensics. In: Proceedings of the 2004 Digital Forensics Research Workshop (2004)

2. Jain, A.K., Murty, M.N., Flynn, P.J.: Data clustering: a review. ACM Comput. Surv. 31, 264–323 (1999)
3. Kohonen, T.: Self-organized formation of topologically correct feature maps. Biological Cybernetics 43(1), 59–69 (1982)
4. Vesanto, J.: Som-based data visualization methods. Intelligent Data Analysis 3(2), 111–126 (1999)
5. Rauber, A., Merkl, D., Dittenbach, M.: The growing hierarchical self-organizing map: Exploratory analysis of high-dimensional data. IEEE Transactions on Neural Networks 13(6), 1331–1341 (2002)
6. Fritzke, B.: Growing grid - a self-organizing network with constant neighborhood range and adaptation strength. Neural Processing Letters 2(5), 9–13 (1995)
7. Dittenbach, M., Rauber, A., Merkl, D.: Recent advances with the growing hierarchical self-organizing map. In: 3rd Workshop on Self-Organising Maps (WSOM), pp. 140–145 (2001)
8. Kruse, W.G., Heiser, J.G.: Computer Forensics: Incident Response Essentials. Addison-Wesley Professional (2001)
9. Pollit, M.M.: Report on digital evidence. In: Proceedings of the Thirteenth International Forensic Science Symposium (2001)
10. Marcella Jr., A., Menendez, D.: Cyber Forensics: A Field Manual for Collecting, Examining, and Preserving Evidence of Computer Crimes, 2nd edn. Auerbach Publications, Boston (2007)
11. Mukkamala, S., Sung, A.H.: Identifying significant features for network forensic analysis using artificial intelligent techniques. Intl. Journal of Digital Evidence 1 (2003)
12. Sommer, P.: Intrusion detection systems as evidence. Comput. Netw. 31, 2477–2487 (1999)
13. Casey, E.: Network traffic as a source of evidence: tool strengths, weaknesses, and future needs. Digital Investigation 1(1), 28–43 (2004)
14. Petersen, J.P.: Forensic examination of log files. Master's thesis, Informatics and Mathematical Modelling, Technical University of Denmark, DTU, Richard Petersens Plads, Building 321, DK-2800 Kgs. Lyngby, Supervised by Assoc. Prof. Robin Sharp (2005)
15. Corey, V., Peterman, C., Shearin, S., Greenberg, M.S., Van Bokkelen, J.: Network forensics analysis. Network forensics analysis. Internet Computing 6(6), 60–66 (2002)
16. Wang, W., Daniels, T.E.: Network forensics analysis with evidence graphs (demo proposal). In: Proceedings of the Digital Forensic Research Workshop (2005)
17. `http://www.guidancesoftware.com`: Guidance software inc. (2005)
18. `http://www.accessdata.com`: Access data corp. (2004)
19. `http://www.foren` sics intl.com: Armor forensics (2006)
20. `http://www.techpathways.com`: Technology pathways, llc. (2004)
21. Beebe, N., Clark, J.: Dealing with terabyte data sets in digital investigations. In: Pollitt, M., Shenoi, S. (eds.) Advances in Digital Forensics. IFIP, vol. 194, pp. 3–16. Springer, Boston (2005)
22. Yin, H.: Data visualisation and manifold mapping using the visom. Neural Networks 15, 1005–1016 (2002)

23. Fei, B., Eloff, J.H.P., Venter, H.S., Olivier, M.S.: Exploring forensic data with self-organizing maps. In: IFIP Int. Conf. Digital Forensics 2005, pp. 113–123 (2005)
24. Fei, B.K.L., Eloff, J.H.P., Olivier, M.S., Venter, H.S.: The use of self-organising maps for anomalous behaviour detection in a digital investigation. Forensic Science International 162(1-3), 33–37 (2006)
25. Ultsch, A., Siemon, H.P.: Kohonen's Self Organizing Feature Maps for Exploratory Data Analysis. In: Proceedings of International Neural Networks Conference (INNC), pp. 305–308. Kluwer Academic Press (1990)

The Application of Evolutionary Heuristics for Solving Soft Security Issues in MANETs

Marcin Seredynski and Pascal Bouvry

Abstract. The goal of this chapter is to demonstrate how the problem of selfish behaviour in mobile ad hoc networks can be solved by means of a soft security approach based on evolutionary game theory. In general, the problem is solved by means of a trust system that allows a node to distinguish between cooperative and selfish network participants, prior to providing them with a forwarding service. Our approach combines a classical trust system with an evolutionary game-theoretic model. The model aims at tuning the parameters of the trust system. We focus on the properties of the trust system that would allow nodes to develop cooperation on the grounds of direct and indirect reciprocity. In particular, we demonstrate, that the numbers of selfish and altruistic nodes have a significant influence on the desired settings of the system.

1 Introduction

The main goal of mobile ad hoc networks (MANETs) is to allow a set of devices (nodes) to wirelessly connect without support of any fixed infrastructure [1]. Such networks are expected to play an important role in civilian and military settings because of their ease of deployment. Packet delivery is based on multi-hop routing, i.e. intermediate nodes convey information on behalf of others. As a result, cooperation in packet forwarding is one of the network's silent requirements. Most of the devices in MANETs rely on batteries, thus one of the network's most important characteristics is the energy-constrained operation [2]. In consequence, it is very likely that

Marcin Seredynski
Interdisciplinary Centre for Security, Reliability and Trust, University of Luxembourg, Luxembourg, Luxembourg
e-mail: marcin.seredynski@uni.lu

Pascal Bouvry
Faculty of Sciences, Technology and Communication, University of Luxembourg, Luxembourg, Luxembourg
e-mail: pascal.bouvry@uni.lu

D.A. Elizondo et al. (Eds.): Computational Intelligence for Privacy and Security, SCI 394, pp. 97–114.
springerlink.com

nodes will choose not to participate to routing duties to conserve their energy. As a result, such networks are very likely to suffer from the non-cooperative behaviour of its participants [3, 4, 5]. Therefore, some mechanisms aiming at encouragement of cooperation and punishment of free-riders are required. Unfortunately, traditional security mechanisms do not solve these cooperation-related problems. Their goal is to detect, prevent, or recover from security attacks [6]. Resources are protected from malicious parties by allowing access to authorised users only. The problem of non-cooperative nodes assumes that network participants are selfish but not malicious. It can be resolved if nodes use *trust and reputation systems* coupled with *response mechanisms*. When such an approach is followed by certain number of network participants, a distributed *cooperation enforcement system* is created. In general, such a system belongs to a class of *social control mechanisms*. These mechanisms are also referred to as *soft security* [7] (in opposite to traditional security mechanisms called hard security). Local trust and reputation systems (used independently by network participants) allow nodes to evaluate how cooperative are other network participants. The role of a response mechanism is to tell whether to forward or discard a packet received from a sender with a given level of cooperation. A trust system supplies a node with a subjective evaluation of the level of cooperation of others. On the other hand, the evaluation provided by a reputation system is based on the assessments made by a group of network participants [8].

Our main focus in this chapter is on data types (hereinafter referred to as *trust data*) used by trust systems to evaluate the level of cooperation of network participants. In the cooperation enforcement systems proposed in the literature trust data are usually classified into either *private* (first-hand observations) or *public* (second-hand observations obtained from third parties). A new classification of data into *personal* and *general* was introduced in [9]. However, an experimental evaluation of the influence of these types on the performance of a trust system was not provided. Personal data refer to the status of packets originated by a node itself, while general data refer to the status of packets sent by other nodes. Both data types can be obtained from first and second-hand observations. Such a distinction is related to *direct* and *indirect reciprocity*. These are two well known generic mechanisms underlying cooperation in long term relations among unrelated individuals [10]. The direct reciprocity-based cooperation can be characterized as "I forward your packets and you will reciprocate in the future by forwarding mine", while indirect reciprocity cooperation as "I forward your packets, and somebody else will indirectly reciprocate by forwarding mine". On the basis of the classification four basic modes of a trust system (differing in the use of trust data) can be defined.

In this chapter we demonstrate how the problem of selecting the right mode of a trust system can be addressed using evolutionary game theory and computer simulation. We show that the nature of the problem is game-theoretic, i.e. there are multiple independent decision makers and several decisions of one user will have an impact on the outcome of other network participants. Moreover, users of the network act solely out of self-interest. Our results demonstrate that the performance of a mode is frequency and density-dependent, i.e. it is influenced by all modes present in the network and by the size of the network. We show that two practical results can be obtained. Firstly, the approach allows us to predict the adaptations of the mode of a trust system that will occur in a given network condition. Secondly, it enables to verify whether the adaptations will lead to a stable state of the network. On the basis of these findings, the parameters of a trust system can be tuned so that its behaviour is optimised to the actual conditions of the network.

This chapter is organized as follows. Sect. 2 discusses related work. In Sect. 3 the mechanism for personal and general data collection is explained. Sect. 4 presents a model of the network used in this work. Sect. 5 introduces our evolutionary approach. Sect. 6 shows a selection of results. In the final section we make some concluding remarks.

2 Background and Related Work

This section provides a brief introduction to evolutionary game theory and describes related work regarding data types for trust evaluation in MANETs.

2.1 Evolutionary Game Theory

Evolutionary game theory was brought to widespread circulation in 1973 by John M. Smith and George R. Price in [11]. The authors demonstrated how game theory can be applied to the behaviour of animals. In general, the evolution by natural selection can be seen as an evolutionary game [12]. The players are the organisms, strategies are represented by heritable phenotypes and payoffs are expressed by fitness. Composition of the community influences fitness of a phenotype, which is similar to game-theoretic situations, where success of a strategy depends on the strategy of the co-player. Classical game theory is regarded as a theory of strategic reasoning, where one player is trying to outguess the moves of a co-player. Evolutionary game theory shifts the main focus to population dynamics - successful strategies are the ones that increase their frequencies among the community. Evolutionary game theory moves away from the assumption of *perfect rationality* in favour of viewing the game as an approximation of an actual interaction [13]. While in classical game theory it is perfect rationality that leads players to select sensible strategies, in evolutionary game theory it is replaced by *natural selection*. It has two games - inner and outer [12]. The former is a classical game in which players obtain payoffs according to their own strategies and the strategies of others. The latter is responsible for the evaluation of strategy frequencies according to payoffs received by players using particular strategies. An evolutionary game models a dynamic process by providing a description of how players belonging to a large population interact with each other in repeated randomly matched games. During this process the players adapt their strategies, therefore, any equilibrium that might arise is a consequence of the adaptation process. *Replicator equation* is the foundation of game dynamics: it models how the distribution of strategies in the population changes over time [14].

The share of players using a given strategy grows at a rate equal to the difference between the average fitness of that strategy and the average fitness of the entire population. The most common solution concept is an equilibrium concept called *evolutionary stable strategy* (ESS) [11]. A strategy is evolutionary stable, if under the assumption that almost everyone in the population is using the ESS, no other alternative strategy can do better than the ESS. Let us define $E_S(A)$ as the expected payoff to S played against A. If for all $A \neq S$ we have $E_S(S) > E_S(A)$ (strict Nash equilibrium), then S is an ESS. In the case when $E_S(S) = E_S(A)$, S is still an ESS if an additional stability condition is satisfied, i.e. $E_A(S) > E_S(S)$. Due to the stability condition the criteria for an ESS are more demanding compared to the Nash equilibrium. If the ESS has a higher expected payoff than the alternative strategy, natural selection will simply eliminate the latter.

Despite its biological context evolutionary game theory has recently become of interest to economists, philosophers, sociologists and computer scientists. A pioneering role was played by Robert Axelrod. In his now-classical book "The Evolution of Cooperation" [15] the author showed how evolutionary game theory can be applied to address a dynamic problem of cooperation among agents who act solely out of self-interest. Classic game theory has already been applied at different protocol layers in MANETs. Its applications range from the problem of selfish behavior of nodes in forwarding packets, interference avoidance to distributed power control (for a survey of the applications see [16]).

2.2 Trust Data for Cooperation Enforcement

The goal of distributed cooperation enforcement mechanisms in MANETs is to create a *strategy-proof* network, i.e. network that is resistant to selfish users wanting to exploit it [17]. Such mechanisms fall into two categories referred to as (i) *pricing* and (ii) *trust-based*. The first (i) approach can be seen as an economic view of the problem as nodes have to pay for receiving a service and are paid for providing it. The trust-based approach (ii) is composed of two elements - a *trust system* and a *response mechanism*. The goal of the former is to collect information about the behaviour of other network participants in order to evaluate their level of cooperation. The latter defines how forwarding requests are processed (depending on how cooperative is the source of the request). The second approach might also include a reputation system.

Several cooperation enforcement mechanisms have been proposed in the literature (see e.g. [18, 19, 20, 21, 22, 23, 24, 25, 26] for trust and reputation-based systems and [3, 27, 28] for pricing-based approaches). A survey of trust and reputation systems in the context of cooperation enforcement in MANETs can be found in [29, 30, 31]. Typically in these systems a distinction is made between the first and the second-hand information. In general, the first-hand observations are more reliable that second-hand observations [32]. The question of whether to use second-hand information or not is basically related to the trade-off between speed of the evaluation of the cooperation level and the robustness of such an evaluation [33]. In the CORE mechanism introduced in [19] the cooperation level is evaluated using first and second-hand evaluations (both having the same significance). However, the second-hand ratings include only information about cooperative behaviour. In consequence, a malicious broadcast of negative ratings for legitimate nodes is avoided. The reliability of the cooperation level evaluation is also positively correlated with the number of evaluations taken into account and its variance. In [18] the authors propose a protocol called *CONFIDANT*, in which the negative second-hand rating is allowed. However, own experience is rated higher than second-hand reports. In [20, 34] the use of the second-hand information is further investigated. A Bayesian approach is introduced: opinions that deviate from the first-hand observations and the opinion of the majority are excluded. As a result, the reputation system is more robust against false accusations and benefits from a faster detection of selfish nodes. In [33] the authors apply the mean-field approach to a proposed stochastic process model to demonstrated that liars have no impact unless their number exceeds a certain threshold. In the SORI algorithm [21], ratings are only exchanged between neighbours. The cooperation level of the node being evaluated is positively correlated with the ratio of packets relayed to packets discarded on behalf of the evaluator. In [35] authors demonstrate that if cooperation is based on indirect reciprocity and a classic watchdog-based mechanism (introduced in [4]) for data collection is used, then discarding packets can be seen as an act of altruistic punishment. In such a situation an intermediate node that drops packets from selfish senders pays a cost expressed in the decrease of its cooperation level among other nodes. If the cost of punishing free-riders is too high then nobody has the incentive to be the punisher.

3 Trust Data Collection and Data Types

Firstly, this section describes the way trust data collection is performed using watchdog-based mechanism. Secondly, the classification of trust data into personal and general is explained.

3.1 Data Collection

Information elements used to derive the level of cooperation of a node are represented by two network events, "packet relayed" and "packet discarded". As a source routing is used, a list of intermediate nodes is included in the packet header. The information regarding the packet forwarding behaviour of other nodes (trust data) is gathered only by nodes that directly participate in a *communication session*. There is no exchange of ratings between nodes. The communication session involves a source node, several forwarders (nodes that are asked to relay packets) and a destination node. Trust data collection is performed in the following way: nodes are equipped with a *watchdog mechanism* (WD) [4] that enables them to check whether the packet has been delivered to the destination. A node that requests another node to relay a packet verifies by means of a *passive acknowledgement* [36] whether the requested node has actually relayed the packet. As an example, let us assume that node S originates a message to node D via intermediate nodes A and B, and eventually the message is discarded by node B (see Fig. 1).

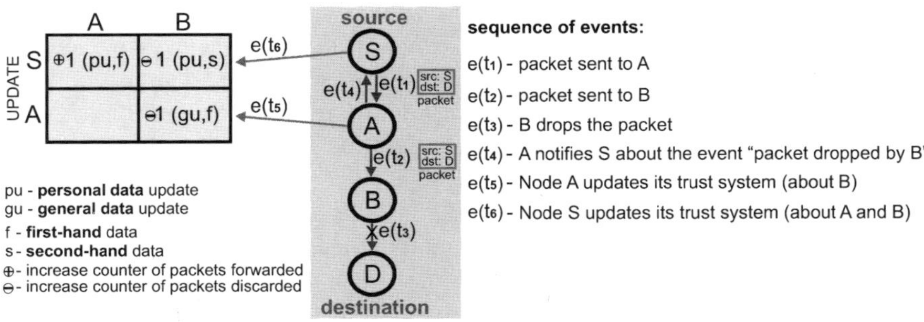

Fig. 1 Example of the WD-based trust data capture mechanism: communication session between nodes S and D failed because the packet was discarded by node B.

This event is recorded by the WD mechanism of node A. Next, this node informs node S about the selfish behaviour of B. As a result, the trust system of node S is updated with two events - "packet relayed by A" and "packet discarded by B", while the trust system of A is updated with the event "packet discarded by B". False accusations are difficult in such a scheme. For a detailed analysis of the performance of the WD mechanism we suggest [4].

3.2 Personal and General Data

Trust data are classified into two types referred to as *personal* and *general* [9]. The former considers status of packets originated by a node itself, while in the latter the status of packets originated by other nodes is taken into account. In consequence, a node in a sender role collects personal trust data, while forwarders collect general data. Both types of data can be obtained from either first and second-hand observations. In the example shown in Fig. 1 node S updates its personal data about nodes A and B, while node A updates its general data about B.

The particular type of data used to evaluate the level of cooperation of the sender (in order to decide whether to forward or drop its packets) results in a development of certain mechanisms underlying cooperation. If such a decision was made on the basis of personal trust data only, cooperation would be build upon direct reciprocity. The mechanism assumes one-on-one interactions between the sender and forwarders. On the other hand, the use of general trust data opens a possibility for a development of cooperation on the basis of indirect reciprocity mechanism.

3.3 Modes of a Trust System

Classification of personal data used in this article allows to create four basic modes of a trust system [9]. In the first mode (denoted by G), a node evaluates the level of cooperation of others using only general trust data. According to the second mode (denoted by P), only personal trust data is used. In the third one (denoted by PG), both data types are used (with the same significance). In the final mode, denoted by PPR, personal data are preferred over general ones, i.e. general data are used only if personal are unavailable.

4 Network Model

In this section we present the model of the network used to evaluate modes of the trust system. First, we describe network assumptions and then we define six types of nodes differing in their forwarding behaviour.

4.1 Network Assumptions

The behaviour of nodes in MANETs is affected by many elements, which include the number and heterogeneity of network participants, monitoring techniques, *a priori* trust relationships, etc. In this work the following assumptions about the network are made:

1. it is self-organising; the network layer is based on a reactive, source routing protocol;
2. topology of the network is unpredictable and changes dynamically, while its size ranges from 60 to 120 nodes;
3. devices are equipped with an omnidirectional antenna with a similar radio range, bi-directional communications and promiscuous mode;
4. *a priori* trust relations do not exist;
5. users might be selfish, but not malicious;
6. decisions about packet forwarding are controlled by dedicated protocols, however users can modify their settings.

4.2 Behaviour of Nodes

The influence of personal and general trust data on the performance of a trust system was evaluated in this work using a simple trust-based forwarding scheme introduced in [35]. Each node uses a local *trust system* and a specific *forwarding approach*. The goal of the former is to evaluate how cooperative the original sender of a packet (source node) is. The latter defines whether the packet should be passed on to the next hop or should be dropped. The *cooperation level* is defined as a ratio between the number of packets forwarded and the number of packets discarded by the source node. The evaluation procedure is as follows: each time an intermediate node i receives a *forwarding request*, it checks the level of cooperation of the original sender s. It uses data collected according to the scheme described in Sect. 3, namely $req\_acc_{s|i}$ and $req\_dsc_{s|i}$, which are numbers of packets forwarded and discarded by s respectively. The notation $s|i$ means that node s is under the evaluation of node i. The cooperation level of node s (a number from the interval $[0,1]$) is calculated as follows:

$$cooperationLevel_{s|i} = \frac{req\_acc_{s|i}}{req\_acc_{s|i} + req\_dsc_{s|i}}.$$

In the next step node i has to decide whether to accept or reject the packet from s:

$$\langle decision \rangle = \begin{cases} forward & \text{if } cooperationLevel_{s|i} >= minCoop_i \\ discard & \text{if } cooperationLevel_{s|i} < minCoop_i, \end{cases}$$

where $minCoop_i$ is a minimal cooperation level required by node i to provide the forwarding service.

The range of data used by i for the evaluation is defined by the mode of its trust system. For instance, if i uses the system in the P mode, then it only verifies whether s forwarded its packets. It means that the behaviour of s towards packets of other nodes is not taken into account. Two cases concerning messages originated by unknown nodes are specified. If the forwarding request occurs in the initial period of the existence of the network (specified by a threshold parameter t_{unkn}), the packet is forwarded with a probability p_1. When the network is established, the packet from an unknown node is forwarded with a probability p_2. In general, p_1 is high (to let newcomers integrate with the network), while p_2 low (to discourage network participants from whitewashing behaviour [37], i.e., changing identity in order to take advantage of the cooperative approach to unknown nodes). Nodes that use such a reciprocity-based behaviour described above are referred to as R-type nodes.

Six types of nodes are defined. The first four types are of R-type, they only differ in the mode of the trust system (P, PPR, PG and G). The remaining types represent two particular patterns of behaviour that one might expect to be present to some extent in a typical MANET. These are *selfish* nodes that forward packets with a probability equal to 0.2 and *altruistic* nodes that unconditionally cooperate (accept all forwarding requests).

Trust data are also used by nodes for sending their own packets to rate the available paths to the destination. The rating is calculated as an arithmetic mean of cooperation levels of all nodes belonging to the route. The path with the best rating is chosen.

5 Evolutionary Approach for the Analysis of Trust Data

Our evolutionary approach is depicted in Fig. 2.

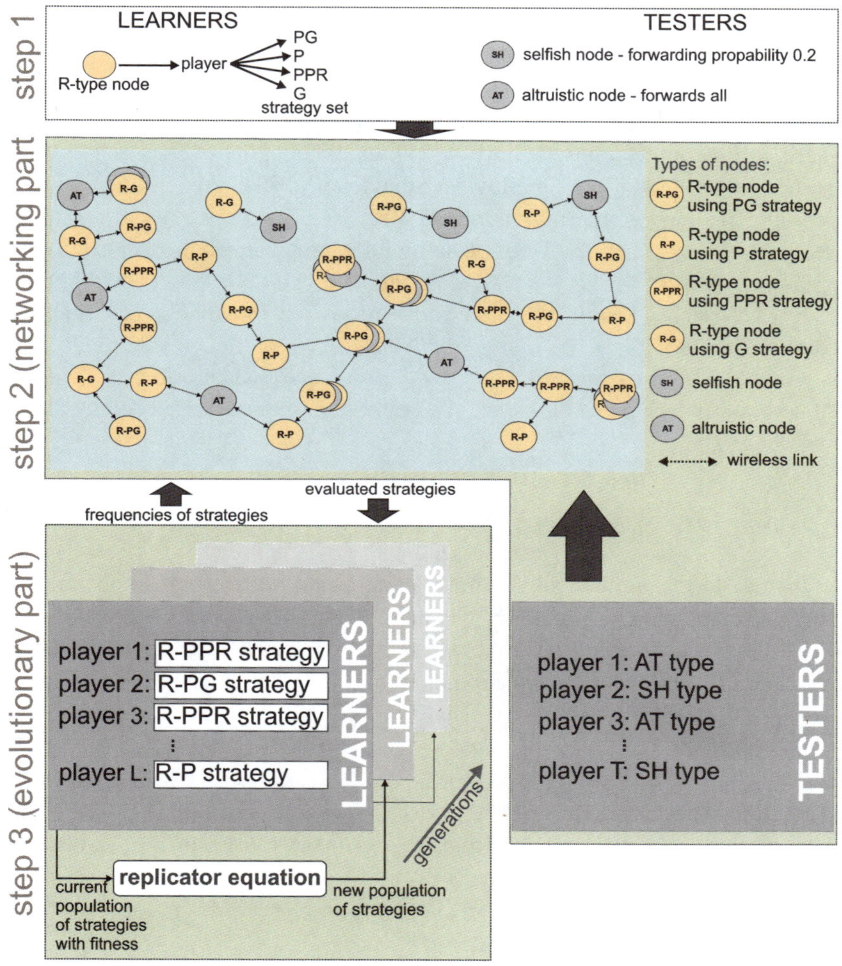

Fig. 2 Overview of the evolutionary approach for the analysis of trust data.

The network is modelled as an evolutionary game, where players correspond to nodes and their strategies are modes of a trust system. Therefore, the strategy set consists of P, PG, PPR and G modes. Games are played between a source of a packet and intermediate nodes that have two choices, either forward or discard the packet. Payoffs (defined by the utility function presented in the next section) obtained by players using a particular strategy are translated into fitness. Strategies are passed through generations. The frequency of a given strategy within the population depends on the average fitness of individuals using that strategy in the preceding generation. The growth rate of a given strategy is described by replicator equations.

Two populations of nodes are defined. Nodes that belong to the first one are referred to as LEARNERS. They are of R-type, thus they can choose one of the four strategies of their trust system. Their goal is to continuously adapt to the network condition by selecting appropriate strategies. The adaptation is modelled by the evolutionary process simulating the survival of the fittest. Therefore, strategies of these nodes converge to a solution (understood as the best mode of a trust system for a given network condition according to a fitness function). Nodes that belong to the second population are referred to as TESTERS. Their goal is to preserve certain properties of the network by employing behaviours (selfishness and altruism) that one could expect to be present in a typical MANET. Therefore, they are composed of selfish and altruistic nodes. Their behaviour does not change through generations. Overall conditions of the network are determined by all nodes, by coevolving population of LEARNERS and fixed (in terms of behaviour) population of TESTERS.

Our evolutionary game has two goals: (i) to discover strategies that will persist through time in a given network condition, (ii) to verify whether the strategies are *evolutionary stable* (once established in a population cannot be invaded by alternative strategies).

5.1 Computer Simulation Procedure

The scheme of the computer simulation is composed of three steps. In the first one, strategies are assigned to LEARNERS. All strategies are represented in equal proportion. In the second step they are evaluated in a MANET. The final step updates the strategies of LEARNERS. Detailed procedure description can be found below.

Step 1: Setup the network.

1. Specify values of the parameters: M as a number of players participating in the network and R as a number of rounds.
2. Setup the population of LEARNERS: specify L as its size and assign the initial strategies to the players.
3. Setup the population of TESTERS: specify T as its size and assign players with types.

Step 2: Evaluate the strategies.

1. Specify r (round number) as $r := 1$.
2. Specify i (source node) as $i := 1$.
3. Randomly select player j (destination of the packet) and intermediate players, forming several possible paths from player i to j.
4. If more than one path is available, calculate the rating of each path and choose the path with the best rating.
5. Let player i initiate a communication session (originate a packet). The packet is next either passed on or dropped by intermediate players according to their forwarding approaches.
6. After the completion of the communication session update trust data.
7. If $i < M$, then choose the next player ($i := i + 1$) and go to step 3. Else go to point 8.

8. If $r < R$, then $r := r + 1$ and go to step 1 (next round). Else, go to point 9.
9. For each strategy calculate the average payoff and stop the evaluation procedure.

Step 3: Update of strategies of LEARNERS using replicator equation.

1. Let x_s^g denote the proportion of players in the population that use strategy s in generation g.
2. The new proportion of players using strategy s in the subsequent generation (x_s^{g+1}) is given by the following equation:

$$x_s^{g+1} = x_s^g \frac{f_s}{\bar{f}}, \tag{1}$$

where f_s is the average fitness of players (described in the next section) that used strategy s and $\bar{f}$ is the average population fitness.

The fitness of a strategy is calculated as an average payoff received by nodes that used the strategy. The payoff of node i is calculated as follows:

$$payoff_i = \frac{nps_i}{npf_i}, \tag{2}$$

where nps_i is a number of packets successfully sent by node i and npf_i is a number of packets forwarded by i.

Steps 2 and 3 (evaluation of strategies and update of their frequencies in the population) are repeated for a predefined number of times (referred to as *generations*).

6 Experiments

The experiments were carried out to study the influence of personal and general data types on the performance of a trust system. Nodes belonging to the population of LEARNERS used one of the four modes of a trust system (introduced in Sect. 3.3). At the beginning of the evolutionary process the four strategies were uniformly distributed among the population of LEARNERS. The performance of these modes was limited to two network sizes. For each size several cases differing in the composition of TESTERS (i.e. in the proportion of altruistic and selfish nodes) were analysed. The main purpose of this chapter was to demonstrate how evolutionary heuristics can be applied to solve the problem of selection of data types, therefore, only a limited set of parameter values was used.

6.1 Network Settings and Parameter Specification

The parameter specifications of the common settings of all experiments and compositions of nodes are given in Table 1. Two cases were analysed. In the first case (referred to as a *small network* case) the network was composed of 60 nodes. In the second one (referred to as a *large network* case), the size was doubled to 120 nodes. One third of the nodes belonged to the population of TESTERS, while the remaining nodes constituted the population of LEARNERS. The initial strategies (P, PPR, PG and P) were uniformly distributed among the population of LEARNERS. However, during the evolutionary process, these proportions changed, as the more successful strategies proliferated, while the weaker ones disappeared from the population. On the other hand, the population of TESTERS did not change over time (constant ratio of selfish to altruistic nodes). The simulation time was set to 600 rounds. The path lengths ranged from 1 up to 5 hops with the following probabilities: one hop - 0.1, two hops - 0.3 and three to five hops - 0.2. The *minCoop* values ranged from 0.85-1 (each value from the interval was chosen with an equal probability). The number of available paths from a source to a given destination ranged from 1 to 4 (with an equal probability). Each experiment (corresponding to a given network setting) was repeated 100 times (referred to as runs). The mean value of the performance measure of nodes was calculated over all runs of the experiment. The experiments were carried out according to the simulation procedure described in Sect. 5.1.

Table 1 Specification of the parameters and settings of nodes.

General parameters	**Value**
simulation time (number of rounds (R))	600
strategy set	P, PG, PPR, G
minCoop	0.85-1 (equiprobable)
forw. prob. of LEARNERS towards unknown (p_1)	1.0
forw. prob. of LEARNERS towards unknown (p_2)	0.3
t_{unkn} of LEARNERS	round # 50
trust of an unknown node in path rating mechanism	0.5
path length/probability of a given # of hops	1/0.1, 2/0.3, 3-5/0.2
number of available paths	1-4 (equiprobable)
Types/popularity of nodes	**small/large network cases**
total number of nodes	60/120
number of TESTERS	40/80
number of LEARNERS	20/40
initial popularity of each strategy (P, PG, PPR, G)	10/20
number of selfish nodes	0-10/0-20
number of altruistic nodes	0-10/0-20

6.2 *Results and Discussion*

The evolutionary outcomes as a function of the number of selfish nodes in the population of TESTERS are shown in Fig. 3a (small network) and Fig. 3b (large network). The x-axis refers to the number of selfish nodes present among TESTERS, which means that the remaining nodes belonging to that population are the altruists. The number of selfish nodes was ranging from 0 to the size of the entire population of TESTER (which is 20 in the small network and 40 in the large network case). The y-axis demonstrates the changes in the frequency of the strategies used by TESTERS. For instance, in the case of a small network without selfish nodes the following changes were observed. The four strategies coexisted until the 44th generation, where the G strategy was eliminated. The PPR strategy survived until generation number 50. Next, the two strategies (PG and P) coevolved until generation number 156. From that generation onwards, the P strategy was the only one that survived in the 100 runs of the experiment corresponding to this setting. This situation is shown in details in Fig. 4a (small network) and Fig. 4b (large network).

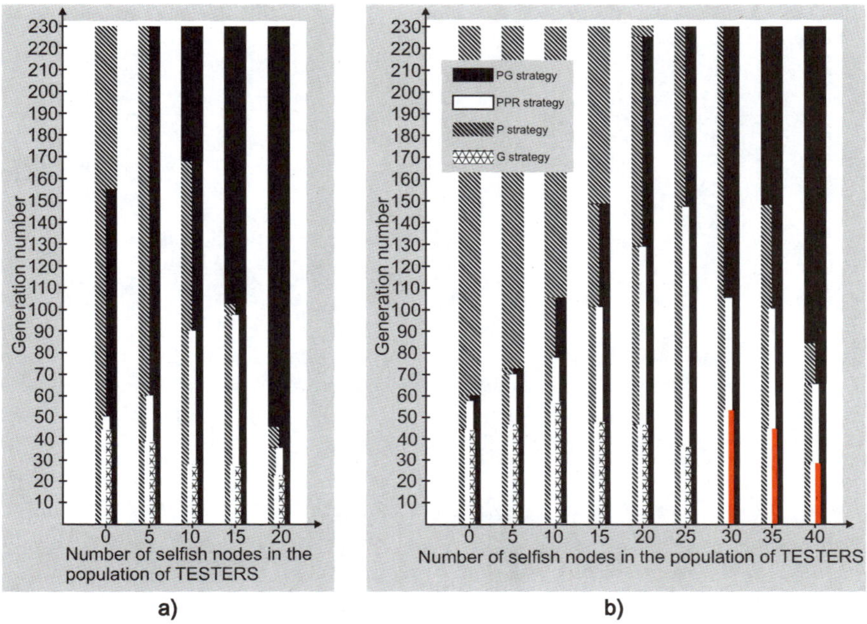

Fig. 3 Evolutionary outcomes of LEARNERS under the continuous replicator dynamics as a function of the number of selfish nodes in the small network (a) and the large network (b). In both cases, a small number of selfish nodes promoted the P strategy, while a large number worked in favour of the PG strategy.

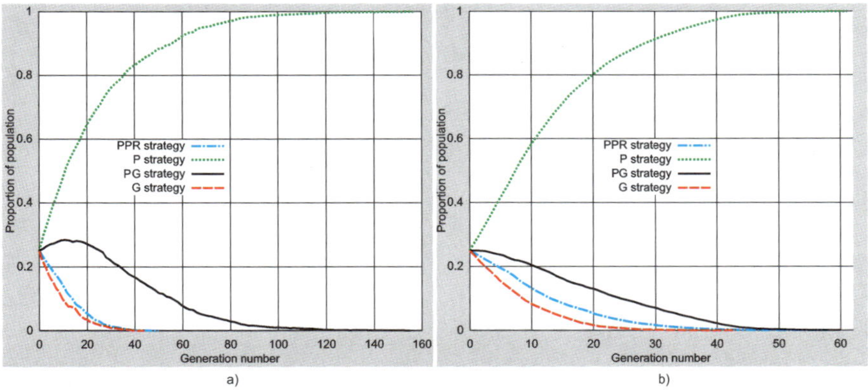

Fig. 4 Evolutionary outcome of LEARNERS under the continuous replicator dynamics in the small network (a) and the large network (b). Selfish nodes were not present in the network. In both cases the final outcome was the P strategy.

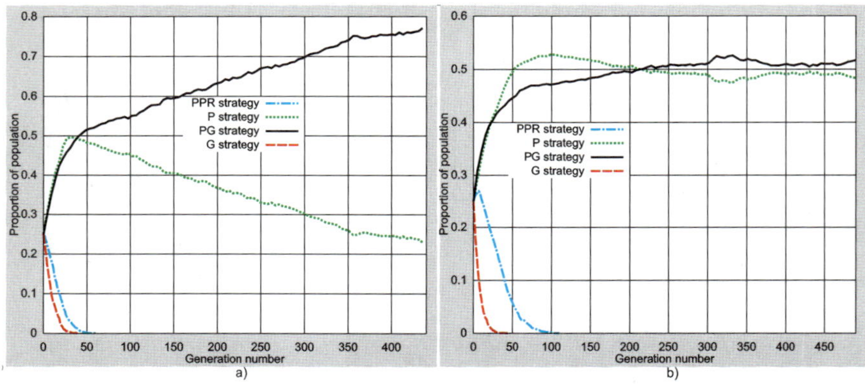

Fig. 5 Evolutionary outcome of LEARNERS under the continuous replicator dynamics in the small network with 5 selfish nodes (a) and the large network with 26 selfish nodes (b). In the small network the outcome was dominated by the PG strategy, while in the large network, PG and P strategies obtained similar performances.

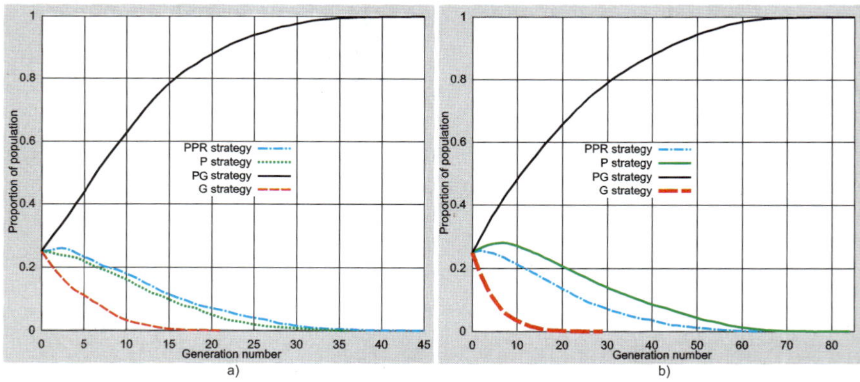

Fig. 6 Evolutionary outcome of LEARNERS under the continuous replicator dynamics in a small network with 20 selfish nodes (a) and a large network with 40 selfish nodes (b). In both cases the final outcome was the PG strategy.

In general, in the presence of a low number of selfish nodes, the P strategy was the unique outcome (in both network sizes). As soon as there were 5 selfish nodes (in the small network) and 26 selfish nodes (in the large network) the outcomes were different. In the case of the small network the P strategy was no longer the winner. It was slowly eliminated by the PG strategy (see Fig. 5a). In the case of the large network, both strategies divided the population between themselves (see Fig. 5b). The next turning point for the small network was observed when at least half of the population of TESTERS (i.e. 10 nodes, which is around 16% of the whole population) was composed of selfish nodes. In such a case, PG strategy was the outcome. A similar observation was made in the large network. This however required almost 90% of TESTERS to be selfish (35 nodes, which is about 29% of the whole network). From that points on, the PG strategy was the unique outcome in the small and the large network cases. The more selfish nodes were present, the easier was the victory of the PG strategy. The extreme case, where all TESTERS were selfish is shown in Fig. 6a (small network) and Fig. 6b (large network). In both cases the PG strategy easily won after around 45 (small network) and 80 generations (large network).

Additional experiments were carried on to tell whether the discovered outcomes were evolutionary stable against the other three strategies. For a given strategy to be evolutionary stable, it has to be the most profitable in a population of nodes, where almost everyone is using that strategy. The procedure was as follows: the majority of the population of LEARNERS (95%) used the strategy that was found to work best in a given network condition (the condition was defined by the network's size and the number of selfish nodes). The strategy will hereinafter be referred to as the *main strategy*. The remaining LEARNERS (hereinafter referred to as *invaders*) used one of the three available alternative strategies. All invaders used the same strategy in a given experiment, therefore, three independent cases were analysed. Table 2 shows the average payoff to each strategy for the small network case and Table 3 shows the results corresponding to the large network case.

Table 2 Average payoffs received by the main strategy and the strategy used by the invaders in the small network case. The main strategy was used by the majority of LEARNERS (38 nodes), while the alternative one was used by the invaders (2 nodes).

# of selfish nodes	the main strategy	PG	PPR	P	G
0	P	0.512 (0.38)	0.479 (0.03)	**0.520-0.524**	0.452 (0.04)
5	P	*0.504 (0.51)*	0.486 (0.08)	**0.503-0.506**	0.442 (0)
5	PG	**0.518-0.522**	0.489 (0.05)	*0.520 (0.52)*	0.447 (0.02)
10	PG	**0.518-0.524**	0.496 (0.10)	0.507 (0.29)	0.427 (0)
15	PG	**0.567-0.558**	0.519 (0.02)	0.520 (0.07)	0.431 (0)
20	PG	**0.674-0.697**	0.559 (0)	0.553 (0)	0.450 (0)

Table 3 Average payoffs received by the main strategy and the strategy used by the invaders in the large network case. The main strategy was used by the majority of LEARNERS (76 nodes), while the alternative one was used by the invaders (4 nodes).

# of selfish nodes	the main strategy	PG	PPR	P	G
0	P	0.492 (0)	0.469 (0)	**0.529-0.532**	0.445 (0)
5	P	0.490 (0.03)	0.470 (0)	**0.519-520**	0.433 (0)
10	P	0.490 (0.08)	0.475 (0)	**0.510-0.512**	0.440 (0)
15	P	0.490 (0.18)	0.477 (0.01)	**0.505-0.507**	0.434 (0)
20	P	0.495 (0.31)	0.48 (0.06)	**0.503-0.505**	0.437 (0)
26	P	*0.508 (0.52)*	0.487 (0.04)	**0.506-0.509**	0.435 (0)
26	PG	**0.504-0.510**	0.492 (0.2)	*0.509 (0.54)*	0.413 (0)
30	PG	**0.516-0.519**	0.497 (0.04)	0.514 (0.39)	0.413 (0)
35	PG	**0.538-0.545**	0.508 (0.05)	0.522 (0.13)	0.413 (0)
40	PG	**0.57-0.58**	0.523 (0.01)	0.537 (0.04)	0.414 (0)

One can notice that, in most cases the previously found evolutionary outcomes turned out to be evolutionary stable. For instance, in the small network without selfish nodes the P strategy was the most successful one (see Fig. 3). Its evolutionary stability is shown in the first row in Table 2. The average fitness was ranging from 0.520 to 0.524 (depending on the alternative strategy used). Fitness values for the alternative strategies were lower than the one of the main one (0.452-0.512). The values in the parenthesis show the ratio of runs won by the alternative strategy against the main one. One can notice, that in the mentioned example, the best performance among the alternative strategies was achieved by the PG strategy. In 38% of runs of the experiment it obtained a higher fitness value than the main strategy. Nevertheless, the P strategy had a greater probability of achieving a better fitness than the PG strategy, therefore, P is an ESS. There are two cases where the discovered strategies were found not to be an ESS: (i) when 5 selfish nodes were present in the small network, and (ii) when 26 selfish nodes were present in the large network. In both cases, P and PG strategies were tested for potential stability. Each time, one of them was the majority strategy, the other one (acting as the alternative strategy) obtained a better fitness and won more than 50% or the runs. It means that neither of the strategies is an ESS.

7 Conclusion

We have demonstrated that for dynamic problems, where one needs to study behaviour of the population of independent agents, very promising tools are provided by the evolutionary game theory. We have addressed a particular problem of cooperation in MANETS related to the data types for trust evaluation. An indirect approach letting nodes to choose the types independently allowed to discover what pattern of behaviour can be obtained by network participants acting out of self-interest. We have shown that by using the model two practical results can be achieved. Firstly, one can verify what is the most profitable strategy for a node in a given network condition. Secondly, one can evaluate whether the strategy is stable, which means that, once used by most network participants, will remain the best choice from a node's point of view. The results obtained by the evolutionary model do not only show how nodes might behave, but they also provide certain insights into the design of trust evaluation mechanisms for MANETs. Ideally these mechanisms should somehow adapt the use of personal and general data according to the network size. We have shown that the choice of an appropriate mode depends on the number of selfish nodes present in the network. If the number is high, the best performance is achieved by nodes that use both personal and general data. In a cooperative network (small number of selfish nodes), the use of only personal data is the best choice for a node.

Acknowledgements. This work has been partially founded by C08/IS/21 TITAN Project (CORE programme) financed by National Research Fund of Luxembourg.

References

1. Jurdak, R.: Wireless Ad Hoc and Sensor Networks: A Cross-Layer Design Perspective. Springer, Heidelberg (2010)
2. Corson, S., Macker, J.: Mobile ad hoc networking (manet): Routing protocol performance issues and evaluation considerations. IETF RFC 2501 (1999), http://www.ietf.org/rfc/rfc2501.txt
3. Buttyan, L., Hubaux, J.P.: Nuglets: a virtual currency to stimulate cooperation in self-organized mobile ad hoc networks, Swiss Federal Institute of Technology, Tech. Rep. DSC/2001/001 (2001)
4. Marti, S., Giuli, T., Lai, K., Baker, M.: Mitigating routing misbehavior in mobile ad hoc networks. In: Proc. ACM/IEEE 6th International Conference on Mobile Computing and Networking (MobiCom 2000), pp. 255–265 (2000)
5. Michiardi, P., Molva, R.: Simulation-based analysis of security exposures in mobile ad hoc networks. In: Proc. European Wireless Conference (2002)
6. Stallings, W.: Cryptography and Network Security: Principles and Practice, vol. 5. Prentice Hall (2010)
7. Rasumsson, L., Jansson, S.: Simulated social control for secure internet commerce. In: Proc. 1996 Workshop on New Security Paradigms (NSPW 1996). ACM (1996)
8. Kwok, Y.-K.: Incentive issues in peer-to-peer systems. In: Bidgoli, H. (ed.) The Handbook of Computer Networks. vol. 3, ch. 46, pp. 168–188. John Wiley and Sons (2007)
9. Seredynski, M., Bouvry, P.: Direct vs. indirect reciprocity trust system in ad hoc networks. In: Proc. 4th IFIP WG 11.11 International Conference on Trust Management IFIPTM 2010, pp. 111–118 (2010)
10. Fehr, E., Gachter, S.: Altruistic punishment in humans. Nature 415, 137–140 (2001)
11. Smith, J.M., Price, G.R.: The logic of animal conflict. Nature 246(5427), 15–18 (1973)
12. Vincent, T.L., Brown, J.S.: Evolutionary Game Theory, Natural Selection, and Darwinian Dynamics. Cambridge University Press (2005)
13. Samuelson, L.: Evolution and game theory. The Journal of Economic Perspectives 16(2), 47–66 (2002)
14. Nowak, M.A.: Evolutionary Dynamics: Exploring the Equations of Life. Belknap Press of Harvard University Press (2006)
15. Axelrod, R.: The Evolution of Cooperation. Basic Books (1984)
16. Srivastava, V., Neel, J., Mackenzie, A., Menon, R., Dasilva, L., Hicks, J., Reed, J., Gilles, R.: Using game theory to analyze wireless ad hoc networks. IEEE Communications Surveys and Tutorials 7(4), 46–56 (2005)
17. Hubaux, J.P., et al.: Cooperation in wireless networks (2011), http://lca.epfl.ch/projects/winet-coop (last checked March 2011)
18. Buchegger, S., Boudec, J.-Y.L.: Performance analysis of the confidant protocol. In: Proc. 3rd International Symposium on Mobile Ad Hoc Networking and Computing (MobiHoc 2002), pp. 226–236 (2002)
19. Michiardi, P., Molva, R.: Core: A collaborative reputation mechanism to enforce node cooperation in mobile ad hoc networks. In: Proc. 6th Conference on Security Communications, and Multimedia (CMS 2002), pp. 107–121 (2002)
20. Buchegger, S., Boudec, J.-Y.L.: The effect of rumor spreading in reputation systems for mobile ad-hoc networks. In: Proc. Workshop on Modeling and Optimization in Mobile, Ad Hoc and Wireless Networks (WiOpt 2003), pp. 131–140 (2003)
21. He, Q., Dapeng, W., Khosla, P.: Sori: a secure and objective reputation-based incentive scheme for ad-hoc networks. In: Proc. Wireless Communications and Networking Conference (WCNC 2004), vol. 2, pp. 825–830 (2004)

22. Buchegger, S., Boudec, J.-Y.L.: Self-policing mobile ad hoc networks by reputation systems. IEEE Communications Magazine, Special Topic on Advances in Self-Organizing Networks 43(7), 101–107 (2005)
23. Yan, L., Hailes, S.: Cooperative packet relaying model for wireless ad hoc networks. In: Proc. 1st ACM International Workshop on Foundations of Wireless Ad Hoc and Sensor Networking and Computing, pp. 93–100. ACM (2008)
24. Milan, F., Jaramillo, J., Srikant, R.: Achieving cooperation in multihop wireless networks of selfish nodes. In: Proc. Workshop on Game Theory for Communications and Networks. ACM (2006)
25. Yan, L., Hailes, S.: Designing incentive packet relaying strategies for wireless ad hoc networks with game theory. In: Wireless Sensor and Actor Networks II, pp. 137–148. Springer, Boston (2008)
26. Seredynski, M., Ignac, T., Bouvry, P.: Probabilistic Packet Relaying in Wireless Mobile ad Hoc Networks. In: Wyrzykowski, R., Dongarra, J., Karczewski, K., Wasniewski, J. (eds.) PPAM 2009. LNCS, vol. 6067, pp. 31–40. Springer, Heidelberg (2010)
27. Buttyan, L., Hubaux, J.P.: Stimulating cooperation in self-organizing mobile ad hoc networks. Mobile Networks and Applications 8(5), 579–592 (2003)
28. Zhong, S., Chen, J., Yang, Y.: Sprite: A simple, cheat-proof, credit-based system for mobile ad-hoc networks. In: Proc. Twenty-Second Annual Joint Conference of the IEEE Computer and Communications Societies (INFOCOM 2003), vol. 3, pp. 1987–1997. IEEE (2003)
29. Hu, J., Burmester, M.: Cooperation in mobile ad hoc networks. In: Guide to Wireless Ad Hoc Networks. ch. 3, pp. 43–57. Springer, Heidelberg (2009)
30. Yu, B.H., Shen, Z., Miao, C., Leung, C., Niyato, D.: A survey of trust and reputation management systems in wireless communications. Proceedings of the IEEE 98(10), 1755–1772 (2010)
31. Cho, J.H., Swami, A., Chen, I.R.: A survey on trust management for mobile ad hoc networks. IEEE Communications Surveys and Tutorials (to appear 2012)
32. Jøsang, A., Ismail, R., Boyd, C.: A survey of trust and reputation systems for online service provision. Decision Support Systems 43(2), 618–644 (2007)
33. Mundinger, J., Boudec, J.-Y.L.: Analysis of a reputation system for mobile ad-hoc networks with liars. Performance Evaluation 65(3-4), 212–226 (2008)
34. Buchegger, S., Boudec, J.-Y.L.: A robust reputation system for p2p and mobile ad-hoc networks. In: Proc. Second Workshop on the Economics of Peer-to-Peer Systems (2004)
35. Seredynski, M., Bouvry, P.: The cost of altruistic punishment in indirect reciprocity-based cooperation in mobile ad hoc networks. In: Proc. Sixth IEEE/IFIP International Symposium on Trusted Computing and Communications (TrustCom 2010), Hong Kong, China, pp. 749–755 (2010)
36. Jubin, J., Turnow, J.D.: The darpa packet radio network protocols. Proceedings of the IEEE 75(1), 21–32 (1987)
37. Feldman, M., Papadimitriou, C., Chuang, J., Stoica, I.: Free-riding and whitewashing in peer-to-peer systems. IEEE Journal on Selected Areas in Communications 24(5), 1010–1019 (2006)

Mixing Genetic Algorithms and V-MDAV to Protect Microdata

Agusti Solanas, Úrsula González-Nicolás, and Antoni Martínez-Ballesté

Abstract. Protecting the privacy of individuals, whose data are released to untrusted parties, is a problem that has captured the attention of the scientific community for years. Several techniques have been proposed to cope with this problem. Amongst these techniques, *microaggregation* is able to provide a good trade-off between information loss and disclosure risk. Thus, many efforts have been devoted to its study. Microaggregation is a statistical disclosure control (SDC) technique that aims at protecting the privacy of individual respondents by aggregating the information of similar respondents, so as to make them undistinguishable. Although microaggregation is a very interesting approach, to microaggregate multivariate data sets optimally is known to be an NP-hard problem. Consequently, the use of heuristics has been suggested as a possible strategy to solve the problem in a reasonable time. Specifically, genetic algorithms (GA) have been shown to be able to find good solutions to the microaggregation problem for small, multivariate data sets. However, due to the very nature of the problem, GA can hardly cope with large, multivariate data sets. With the aim to apply them to large data sets, those have to be previously partitioned into smaller disjoint subsets that the GA can handle separately. In this chapter, we summarise several proposals for partitioning data sets, in order to apply GA to microaggregate them. In addition, we elaborate on the study of a partitioning strategy based on the variable-MDAV algorithm, we study the effect of several parameters, namely the dimension, the aggregation parameter (k), the size of the data sets, etc. Also, we compare it with the most relevant previous proposals.

Agusti Solanas · Úrsula González-Nicolás · Antoni Martínez-Ballesté
UNESCO Chair in Data Privacy in the Department of Computer Engineering and Mathematics, Universitat Rovira i Virgili, Av. Paisos Catalans 26, 43007, Tarragona, Catalonia, Spain
phone: +34 977 558867
e-mail: `agusti.solanas@urv.cat`

D.A. Elizondo et al. (Eds.): Computational Intelligence for Privacy and Security, SCI 394, pp. 115–133.
springerlink.com

1 Introduction

We live in the information era[1] in which we use the Internet to search information about books, music, friends, and almost every imaginable subject. At the same time, we share tonnes of information about ourselves – some times consciously and some others unconsciously – with thousands of people and machines. Statistical agencies, private companies and Internet search engines are just a few examples of entities that daily collect data from many people. Hence, our habits, tastes, and hobbies might be inferred at astonishingly high speeds. Nowadays it is easier than ever to collect personal data from friends, colleagues and even strangers. The information is only a few clicks away from us.

Most countries have legislation which compels national statistical agencies to guarantee statistical confidentiality when they release data collected from citizens or companies; see [10] for regulations in the European Union, [2] for regulations in Canada, and [27] for regulations in the United States. Thus, protecting individual privacy is a key issue for many institutions, namely statistical agencies, Internet companies, manufacturers, etc.; and many efforts have been devoted to develop techniques that guarantee some degree of individual privacy. Many fields of science are involved in the development of privacy preserving methods: cryptography, statistics and artificial intelligence are just a few examples. Notwithstanding, the field of statistical disclosure control (SDC) was the one in which the problem was formally born; firstly tabular data were considered, and more recently microdata (*i.e.* the personal data collected from individual respondents) have become the focus of attention. In this chapter we concentrate on the protection of microdata.

With the aim to protect individuals, microdata sets (containing personal information about individuals) must be properly modified prior to their publication. The degree of modification can vary between two extremes: (i) encrypting the microdata and, (ii) leaving the microdata intact (i.e. with no modifications). Fig. 1 illustrates the effect of modifications over the data utility and confidentiality. In one extreme, the protection is perfect (*i.e.* only the owner of a secret key can see the data), however, the utility of the data is practically zero because the encrypted microdata can hardly be studied or analysed. In the other extreme, the microdata are extremely useful (*i.e.* all their information remains available), however, the privacy of the respondents is endangered. SDC methods for microdata protection aim at distorting the original data set to protect respondents from being identified whilst maintaining, as much as possible, some of the statistical properties of the data and minimising the information loss. The goal is to find the right balance between data utility and confidentiality.

Generally, microdata sets are organised in records that refer to individual respondents. Each record has several attributes. Table 1 is an example [2] of a microdata set with 8 records and 6 attributes that contain information about the *"Name"* of the

[1] Also known as the *Computer Age* and the *Information Age*.

[2] The social security numbers in this table are invented, any resemblance to real people, living or dead is purely coincidental.

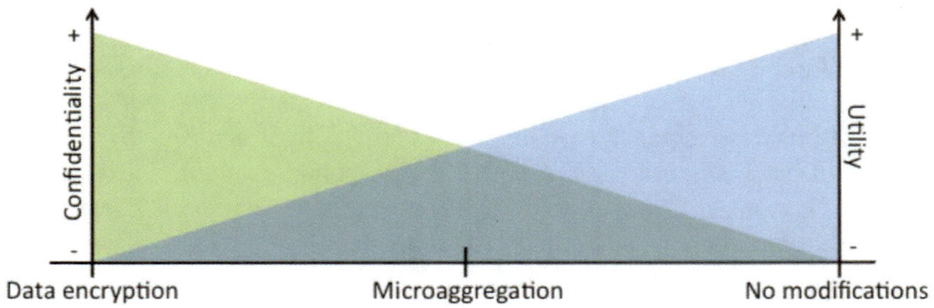

Fig. 1 Relation between data utility and data confidentiality. Microaggregation aims at finding a good balance between data utility and data confidentiality.

respondent, his/her *"Social security number"*, the *"City"* where he/she lives, his/her *"Weight"*, the *"Job"* he/she has, and whether he/she has *"Cancer"*.

More generally the attributes that can appear in a microdata set **X** can be classified in three categories as follows:

1. *Identifiers*. Attributes in **X** that *unambiguously* identify the respondent. For example, passport numbers, full names, etc. In our example (*cf.* Table 1) the attributes *"Name"* and *"Social security number"* are identifiers.
2. *Key attributes*. Those in **X** that, when properly combined, can be linked with external information sources to re-identify some of the respondents to whom some of the records refer. For example, address, age, gender, etc. In our example (*cf.* Table 1) the attributes *"City"*, *"Weight"* and *"Job"* are key attributes.
3. *Confidential/Outcome attributes*. Those containing sensitive information on the respondent, namely salary, religion, political affiliation, health condition, etc. In our example (*cf.* Table 1) the attribute *"Cancer"* is a confidential attribute.

Due to the fact that the re-identification of respondents is to be avoided, we can assume that the identifiers in **X** are removed or encrypted before the microdata set is released. However, unlike identifiers, key attributes and confidential attributes cannot be removed from **X** without significantly degrading the quality of the information.

Table 1 Toy example of a microdata set.

Row	Name	SS. Number	City	Weight	Job	Cancer
1	John F.	39727383	BigCity	76	Baker	NO
2	Alice C.	74837282	Village	89	None	NO
3	Mary Q.	73617282	Hamlet	56	Doctor	YES
4	John M.	88837262	Village	79	None	YES
5	Frank P.	19273837	Village	82	Farmer	NO
6	Marge S.	87483929	Village	68	Nurse	NO
7	Homer S.	19284646	BigCity	110	None	NO
8	Tony M.	74837272	BigCity	78	Baker	NO
...	...	...	...	...	...	...

Unfortunately, removing or encrypting identifiers does not provide enough protection against re-identification to respondents. To make this point clear let us consider our example in Table 1. Imagine that we have removed the "name" and the "social security number" of each respondent (*i.e.* the identifiers), so an external observer only knows the "City" where respondents live, their "Job", their "Weight" and whether they have cancer. In this scenario, it is quite apparent that respondent number 3 can be easily re-identified because the "doctor" in a small hamlet is generally a well-known person and, in most cases, there is only one doctor in hamlets.

From the example above, it might become clear that privately releasing microdata is not so straightforward and requires procedures far more sophisticated than simply removing identifiers. In the field of SDC, several techniques have been proposed to cope with this problem, namely noise addition, rank swapping, microaggregation, etc. Each of these techniques has its pros and cons. Microaggregation is the most recent of these, and has shown a natural capability of preserving interesting properties such as k-anonymity. Thus, in this chapter we will concentrate on microaggregation.

1.1 Contribution and Plan of the Chapter

In the following lines we focus on microaggregation and we show how to improve its results, in terms of information loss, by means of mixing the simplicity of classical microaggregation with the great exploration capabilities of genetic algorithms. We recall and analyse a two-step partitioning method based on the Variable-MDAV algorithm that, when properly combined with genetic algorithms, leads to clear improvements with respect to classical microaggregation.

The rest of the chapter is organised as follows: Section 2 summarises some fundamental concepts of microaggregation and some previous work on the use of genetic algorithms to improve the results of classical microaggregation. Section 3 elaborates on some partitioning methods for microaggregation with genetic algorithms and recalls a proposal based on the V-MDAV algorithm. In Section 4, we present the experimental results obtained over synthetic and real data sets. Finally, the chapter concludes in Section 5 with some final remarks.

2 Background and Related Work

In this section we elaborate on the microaggregation problem and we show how microaggregation protects the privacy of individuals by averting re-identification. We also introduce our previous work on genetic algorithms applied to protect individuals' privacy.

2.1 Microaggregation and Related Techniques

Microaggregation is a family of SDC methods based on data perturbation. In these methods, the microdata set is modified (distorted) prior to its publication. Microaggregation can be operationally defined in terms of two steps:

- *Partition*. The first step consists in partitioning the data set into several parts (groups, clusters) in such a way that the records in the same part are as much similar as possible (*i.e.* within-part homogeneity is maximised). Moreover, the cardinality of each part must be greater than a predefined microaggregation parameter k. A partition meeting such cardinality requirements is known as k-partition [5].
- *Aggregation*. Once the k-partition is obtained, an aggregation operation (usually the mean) is used to compute the centroid of each part. Then, each record in a part is replaced by the part's centroid, that is the average record. Hence, the perturbed microdata contain sets of k indistinguishable records.

The second step clearly aims at guaranteeing k-anonymity to respondents: a data set is said to satisfy k-anonymity for $k > 1$ if, for each combination of values of key attributes, at least k records exist in the data set sharing that combination [20]. K-anonymity demands that every tuple in the microdata table released be indistinguishably related to no fewer than k respondents.

Microaggregation was originally designed for continuous numerical microdata. Since one can perform arithmetical operations on continuous numerical data, the average can be used to compute the centroid. However, if attributes are categorical (*i.e.* they take values over a finite set) or nominal, standard arithmetical operations are not feasible. Hence, other techniques must be designed for categorical and nominal attributes [26, 7].

As stated in the introduction, privately releasing microdata is not straightforward, and requires other more sophisticated techniques than merely removing identifiers. Note that in Table 1, if the "Name" and "Social security number" are removed from the microdata set, one can still identify the only doctor in Hamlet and, thus, infer his/her value for the outcome attribute. Moreover, if only one person in "Village" weights 79 kg and is currently unemployed, he/she could be easily identified. Hence, to reduce the risk of re-identification, several attributes must be deleted, encrypted or perturbed.

Table 2 shows a protected version of our toy example of microdata. The key attributes (i.e. Name and SS. Number) have been removed. The attribute "City" is a

Table 2 Toy example of protected microdata. Key attributes (Name and SS. Number) have been removed; Weight is a numerical attribute that has been microaggregated with ($k = 3$); City is a categorical attribute that has been generalised by the state it belongs to.

Row	City	Weight	Job	Cancer
1	Northstate	67	Baker	NO
2	Eaststate	99	None	NO
3	Northstate	67	Doctor	YES
4	Eaststate	79	None	YES
5	Eaststate	79	Farmer	NO
6	Eaststate	67	Nurse	NO
7	Northstate	99	None	NO
8	Northstate	79	Baker	NO
...	...	...	...	...

categorical attribute that has been generalised using the state to which the city belongs (e.g. BigCity and Hamlet belong to Northstate and Village belongs to Eaststate). Finally, "Weight" is a numerical attribute and so it has been microaggregated with a privacy parameter $k = 3$. Clearly, releasing the protected microdata of our toy example protects the privacy of the respondents. Note that the doctor of Hamlet may be indistinguishable from the other respondents working as doctors in the same state.

2.1.1 The Quality of Microaggregation

According to [5] the main problem in SDC using perturbative methods is to provide sufficient protection without seriously damaging the information contained in the original data. For instance, the statistics computed on the perturbed data set should not differ significantly from the statistics that would be computed from the original data set. In a nutshell, the goal is to hide respondents amongst similar ones (*i.e.* a low disclosure risk) while minimising the loss of information. In our toy example, the average value for the attribute "Weight" before microaggregation is 79.75, and after microaggregation, the average is quite similar (79.50).

The *homogeneity* of the parts formed in the first step of microaggregation plays a key role in the quality of the perturbed data: the higher the homogeneity, the lower the information loss. In that sense, the within-part errors *Sum of Square Errors* (SSE) is a common homogeneity measure [9, 28, 17]. This is the sum of squared Euclidean distances from the average record of each part to every record in the part. For a given k-partition, the SSE is computed as shown in Expression 1:

$$\mathbf{SSE} = \sum_{i=1}^{s}\sum_{j=1}^{n_i}(x_1^{ij} - \bar{x}_1^i, x_2^{ij} - \bar{x}_2^i, \ldots, x_d^{ij} - \bar{x}_d^i) \times \begin{pmatrix} x_1^{ij} - \bar{x}_1^i \\ x_2^{ij} - \bar{x}_2^i \\ \vdots \\ x_d^{ij} - \bar{x}_d^i \end{pmatrix}, \tag{1}$$

where s is the number of parts in the k-partition, n_i is the number of records in the i-th part, x_p^{ij} is the value of the p-th attribute of the j-th record in the i-th part, $\bar{x}_p^i$ is the value of the p-th component of the centroid of the i-th part, and d is the number of attributes/components of the microdata set (*i.e.* the dimension of each record). This expression could be written more compactly by using vector notation as shown in Expression 2:

$$\mathbf{SSE} = \sum_{i=1}^{s}\sum_{j=1}^{n_i}(\mathbf{x}_{ij} - \bar{\mathbf{x}}_i)(\mathbf{x}_{ij} - \bar{\mathbf{x}}_i)^T, \tag{2}$$

where s is the number of parts in the k-partition, n_i is the number of records in the i-th part, $\mathbf{x}_{ij}$ is the j-th record in the i-th part, and $\bar{\mathbf{x}}_i$ is the centroid of the i-th part.

Given a microdata set $\mathbf{X}$, the optimal k-partition is the one with minimum SSE. In [12] Hansen and Mukherjee show that the optimal k-partition of a univariate data set can be found in polynomial time. In contrast, finding the optimal k-partition of a multivariate data set is known to be an NP-hard problem [19]. Due to the hardness of the problem, a plethora of microaggregation heuristics have been proposed. The main aim of these heuristic methods is to obtain k-partitions with low values of the SSE, in reasonable times (*i.e.* with a feasible computational cost).

2.1.2 Heuristics for Microaggregation

Multivariate microaggregation heuristics can be classified into two categories depending on the size of the generated subsets:

- *Fixed-size heuristics*. These heuristics yield k-partitions where all the parts have size k. Note that if the number of records in the original data set is not a multiple of k, there will be a part with a cardinality between k and $2k-1$.
- *Variable-size heuristics*. These heuristics yield k-partitions where all parts have sizes varying between k and $2k-1$.

In general, fixed-size microaggregation heuristics are computationally efficient since they are simple. The first method for univariate fixed-size microaggregation is presented in [3]. In [5], Domingo-Ferrer and Mateo-Sanz proposed a fixed-size multivariate heuristic for numerical continuous microdata. That method consists in taking the two most distant ungrouped records in the microdata set (the Euclidean distance is used), so as to built two groups at each iteration, using the selected records and their nearest $k-1$ records. This heuristic was substantially improved and their authors implemented a new version for the μ*Argus* SDC software package [14][3]. This heuristic is known as Maximum Distance to Average Vector (MDAV) and improves the efficiency of its predecessor in terms of computational complexity. MDAV is detailed later in this chapter. Some variations of this technique have been published in [16] and [24]. However, in the past few years literature has focused on variable-size heuristics.

Variable-size heuristics can often achieve lower information loss because they are able to adapt the choice of group sizes to the structure of the data set. However, these heuristics suffer from higher computation costs. The idea behind variable group size heuristics is to avoid unnatural k-partitions. Fig. 2(a) shows a bivariate microdata set with nine records microaggregated with $k=3$ and using a fixed-size heuristic. This yields a low within-part homogeneity group (group number 1) and the obtained k-partition is far from optimal. If a variable-size heuristic was used, see Fig. 2(b), the cardinality of the parts varies between k and $2k-1$ and, consequently, the result meets the requirement of minimising the SSE.

In [12], the authors describe a method for optimal univariate microaggregation in polynomial time. This method is used by Domingo-Ferrer *et al.* [4] in order to microaggregate multivariate data sets, by means of ranking the records using several techniques. In [24], Solanas *et al.* add a group extension procedure to a fixed-size heuristic so as to obtain variable-size k-partitions. This technique, entitled Variable MDAV (V-MDAV), is detailed later in this chapter. Other variable-size heuristics can be found in [21, 16, 11].

[3] In this improved version, the centroid of the data set is used as reference for finding initial group records.

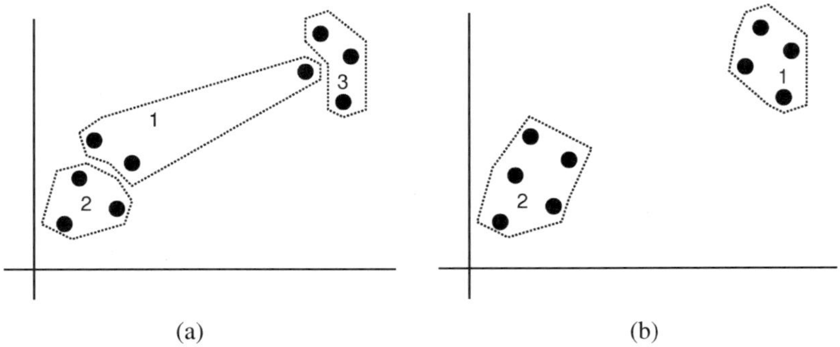

Fig. 2 Split of natural subsets due to the lack of flexibility of fixed-size heuristics. (a) A bivariate microdata set partitioned using $k = 3$ and a fixed-size heuristic; (b) The same data set partitioned using a variable-size heuristic.

2.2 *Microaggregation with Genetic Algorithms*

Genetic algorithms (GA) are biologically inspired optimisers based on the principles of evolution and natural selection [13]. Given a population of individuals (*i.e.* possible solutions), an environment (*i.e.* the problem space), a measure of adaptation to that environment (*i.e.* a fitness function), and a selection and recombination functions, GA look for solutions by means of mutating and recombining individuals that progressively evolve to better fit the environment.

GA have been widely used in a variety of fields. In[5] the authors present a univariate microaggregation approach by means of GA. Moreover, Solanas *et al.* proposed the use of GA to microaggregate small multivariate data sets of up to a hundred records in [25]. A novel N-ary coding was proposed to cope with the multivariate nature of microaggregation and a complete set of experiments was performed to determine the best values for the main parameters of the GA, namely the population size, the crossover rate, the mutation rate, etc.) The GA described in [25] uses the *roulette wheel* selection algorithm, which is based on selecting chromosomes proportionally to their fitness, which was computed as shown in Expression 3:

$$F = \frac{1}{\mathbf{SSE} + 1}\,, \tag{3}$$

where the SSE is computed using Expression 1. There is a variety of selection functions, however, in this chapter we will mainly consider the fitness uniform selection scheme (FUSS) [15].

Although the ideas proposed in [25] were a clear advance in the search for optimal solutions for multivariate microaggregation, they were only applicable to small data sets. Table 3 shows the results of applying the GA over randomly generated data sets: the average value of the SSE obtained using GA is always lower than the SSE obtained using the MDAV heuristic, except for data sets with 100 records

Table 3 Average SSE results of the GA: increasing the number of records [25]. Mutation and crossover rates used are denoted as M_r and C_r respectively.

Data set	MDAV	GA_A	M_r	C_r
20x2	5.39	4.61	0.1	0.7
35x2	5.01	3.59	0.1	0.5
50x2	5.92	4.67	0.1	0.5
100x2	5.48	23.78	0.1	0.0

Table 4 Average SSE results of the GA: increasing the dimensionality [25] (5,7,10 dimensions).

Data set	MDAV	GA
11x5	39.11	29.36
11x7	52.47	43.28
11x10	71.14	68.66

or more. Moreover, as shown in Table 4, the use of GA outperforms MDAV when increasing the dimensionality of the data sets.

In [18] Martínez-Ballesté et al. presented a preliminary study on how to apply GA to medium-sized data sets. The method they proposed consists of two steps: (i) partition the original data set into rough subsets by using a fixed-size microaggregation heuristic and (ii) apply the GA on the obtained rough subsets to generate the microaggregated data set.

The main problem of [18] is that the partitioning method proposed does not consider the natural distribution of the records in the data set, and due to the very nature of fixed-sized microaggregation heuristics, naturally grouped records could be split in different subsets.

3 Our Variable-Size Microaggregation-Based Partitioning Strategy

As stated in the previous section, genetic algorithms can obtain good microaggregated sets when they are applied to small data sets. Notwithstanding, the problem is how to apply them to large data sets. The approach that we propose is the partition of the original large data sets into smaller subsets that the GA can handle, and then, use the GA to obtain a microaggregated data set.

Unfortunately, it is not straightforward to determine how to partition a microdata set in order to obtain subsets that are suitable for a GA-based optimization. As we have explained in the previous section, in [18] Martínez-Ballesté et al. used a fixed-size microaggregation technique to partition a microdata set into smaller parts that the GA described in [25] can handle. Specifically in [18] the authors used the well-known Maximum Distance to Average Vector (MDAV) method [14][22]. In this section, we recall our approach based on a variable-size microaggregation technique, that is, Variable Maximum Distance to Average Vector (V-MDAV) [23][24], and we elaborate on the different ways of applying our solution (*i.e.* one-step partitioning, and two-step partitioning).

3.1 One-Step V-MDAV-Based Partitioning

We know from [25], that a properly tuned GA can deal with up to 50 records with remarkable results. Thus, similarly to [18], our approach consists in generating subsets of cardinality smaller than 50. To do so, we propose the utilisation of the V-MDAV method. In general, V-MDAV improves the results of MDAV by means of adapting to the natural distribution of the records in a microdata set. The V-MDAV works as follows:

1. Compute the distance matrix between all records.
2. Compute the centroid c of the data set.
3. Find the most distant unassigned record r from c. Build a subset around this record, formed by r and its $k-1$ closest records.
4. Extend the subset, as will be explained later.
5. Continue to step 3 until the number of remaining records is smaller than $2k$.
6. If the number of remaining records is smaller than k, assign each remaining record to its closest subset. Otherwise, form the last subset with the remaining records.

The key point of V-MDAV, which makes it better and more powerful than MDAV, is the subset extension step (*i.e.* step 4 of the algorithm). If we remove this step, V-MDAV would behave almost the same as MDAV. Notwithstanding, the extension step allows V-MDAV to adapt to the natural distribution of the records. After generating a subset of k records, the extension step finds possible candidate records, which could join the subset and, if any of these candidate records are closer to the subset than to other unassigned records, they are added to the subset. The extension step works as follows:

1. Find the closest unassigned record u to the lately generated subset.
2. Let d_{in} be the distance from u to the closest record e_i in the lately generated subset.
3. Let u_{out} be the unassigned record closest to u. Let d_{out} be the distance from u_{out} to u.
4. If $d_{in} < \gamma \cdot d_{out}$ and the number of records in the subset is smaller than $2k-1$, add record u to the subset. Return to step 1.
5. Otherwise finish the extension of the subset.

The parameter γ that multiplies d_{out} in step 4 is a *gain factor* that must be tuned depending on the distribution of the records in the microdata set. Determining the best value of γ is not straightforward and is out of the scope of this chapter.

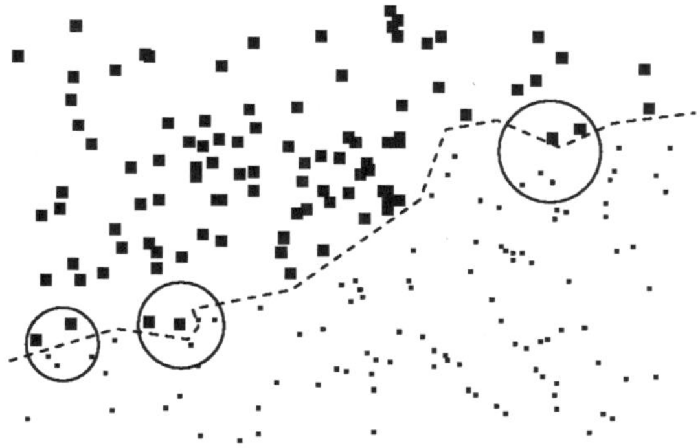

Fig. 3 One-step partitioning. Circles show the undesired breakage of natural groups.

With the aim to illustrate the behaviour of V-MDAV as a partitioning method, we have generated a small bivariate data set consisting of two hundred records and we have partitioned it using V-MDAV with $k = 90$ (*i.e.* two subsets are to be obtained). Fig. 3 shows the partition obtained by V-MDAV. It can be observed that some natural subsets are split and their records are assigned to different parts (*e.g.* see the circles in Fig. 3).

Due to the fact that natural subsets are split in different parts, the GA that we want to apply to each part cannot fully optimise the microdata set in terms of SSE. The breakage of natural subsets is due to the cardinality constraints of microaggregation methods (*i.e.* the generated parts must have at least k records and less than $2k$ records). To overcome this limitation, a two-step partitioning strategy is proposed in the next section.

3.2 Two-Step V-MDAV-Based Partitioning

Instead of partitioning a microdata set in rough parts in a single step (*cf.* previous section), the two-step partitioning proceeds as follows:

1. First step
   - V-MDAV is used to partition the original microdata set $\mathbf{X}$ using a small value of k, usually 3, 4 or 5. Let us call this value k_1.
   - A new data set $\bar{\mathbf{X}}$ is generated and each part of $\mathbf{X}$ is represented by its centroid. Note that, assuming that the original microdata set has n records, the data set $\bar{\mathbf{X}}$ generated at this point has n/k_1 records.
2. Second step
   - V-MDAV is used to partition the data set $\bar{\mathbf{X}}$ generated in the previous step. This time a greater value of k is used, usually 10, 15, 20, 25, 30. Let us call this value k_2.

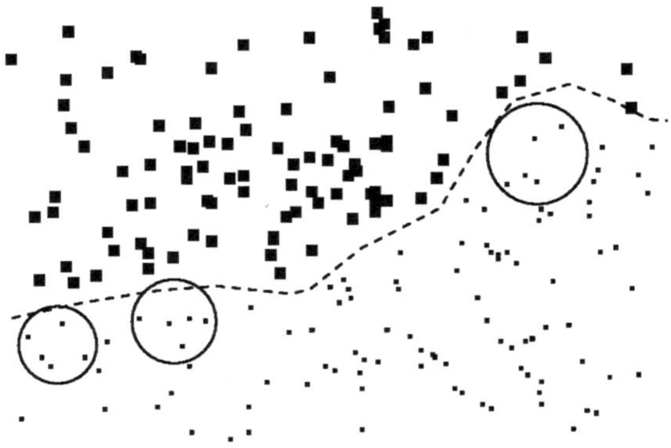

Fig. 4 Two-step partitioning. Circles show the groups that the one-step partitioning technique broke, whilst the two-step technique does not.

- Once the partition of $\bar{\mathbf{X}}$ is finished, each record in $\bar{\mathbf{X}}$ is replaced by the original records that it represents, thus obtaining a partition of the original microdata set with a privacy parameter $k = k_1 \cdot k_2$. Note that due to very nature of V-MDAV, the cardinality of the parts will vary between k and $2k-1$.

By following this procedure, natural subsets of cardinality k_1 are not split. Consequently, the GA that we will apply over each part will have more chances to optimise the microaggregation in terms of SSE. Fig. 4 illustrates the behaviour of the two-step V-MDAV partitioning strategy. It can be observed that subsets that were split with the one-step method are preserved now. By construction it can be assured that no natural subsets of cardinality smaller than k_1 are going to be split. This property, however, cannot be guaranteed when the one-step method is used.

4 Experimental Results

In this section we describe the tests that we have carried out to assess the usefulness of our proposal. Regardless of the partitioning method used (*i.e.* one-step V-MDAV partitioning or two-sept V-MDAV partitioning), we use GA to individually microaggregate each subset [4]. In [23] the roulette wheel selection algorithm [25] was compared to the fitness uniform selection scheme (FUSS) [15], and it was shown that the latter preserves the information better whilst achieving the same levels of privacy. Thus, in this chapter we only use FUSS as a selection strategy. According to the results obtained in [25], we have tuned our GA with the following parameters: mutation rate = 0.1, crossover rate = 0.4, population size = 100 chromosomes and, number of iterations = 5,000.

[4] Note that the microaggregation of all subsets leads to the microaggregation of the complete microdata set.

Table 5 Summary of the experimental results ($L = \frac{SSE}{SST}$). The best result for each data set is highlighted and coloured.

Data set	Classical methods		Hybrid methods			
	MDAV	V-MDAV	MDAV 1 step	V-MDAV 1 step	MDAV 2 steps	V-MDAV 2 steps
Census	0.056619	0.056619	0.059196	0.05927	0.054232	**0.053717**
EIA	0.011017	0.010975	0.012936	0.012969	**0.009542**	0.009834
Tarragona	0.169507	0.158477	0.157482	0.157445	0.159509	**0.151701**
Clustered 1000x2	0.001978	0.000374	0.000571	0.000107	0.000963	**0.000088**
Clustered 1000x3	0.007195	0.000495	0.003088	0.000364	0.004415	**0.00024**
Scattered 1000x2	0.002234	0.002141	0.002399	0.002124	0.002034	**0.001868**
Scattered 1000x3	0.013171	0.013082	0.014019	0.014092	0.012306	**0.012127**

We have used three real microdata sets, which have been widely used as reference microdata sets during the CASC project [1] and in [5, 6, 8, 16]. The microdata sets are the following:

- Census: A real microdata set that contains 1,080 records with 13 numerical attributes (from the US Census Bureau).
- EIA: A real microdata set that contains 4,092 records with 12 numerical attributes (from the US Energy Information Authority).
- Tarragona: A real microdata set containing 834 records with 13 numerical attributes (from the Tarragona Chamber).

In addition we have generated some synthetic data sets to better analyse the behaviour of our proposal depending on the distribution of the records in the data set:

- Clustered1000x2: This is a microdata set with 1000 records and two attributes. The records have been generated so as to naturally cluster in subsets of 3, 4 and 5 records.
- Clustered1000x3: Like the Clustered1000x2 but with 3 attributes.
- Scattered1000x2: This is a microdata set with 1000 records randomly distributed. By construction, these records are not naturally clustered in subsets.
- Scattered1000x3: Like the Scattered1000x2 but with 3 attributes.

We have microaggregated each microdata set by using the different methods explained in this chapter, namely classical MDAV, classical V-MDAV, Hybrid MDAV and Hybrid V-MDAV. For the hybrid methods, we have considered the partition in one step and in two steps.

We have microaggregated all data sets to guarantee 3-anonymity, that is, the results of all the studied methods are microaggregated data sets where each record has, at least, 2 other indistinguishable records. To do so, we have fixed the privacy parameter $k = 3$. After microaggregating the data sets, we have analysed the information loss introduced by each method. The information loss has been measured by using the following expression:

$$\mathbf{L} = \frac{\mathbf{SSE}}{\mathbf{SST}}, \tag{4}$$

where the SSE is computed using Expression 1 and the total sum of squares (SST) is computed as follows:

$$\mathbf{SST} = \sum_{i=1}^{n} (\mathbf{x}_i - \bar{\mathbf{x}})(\mathbf{x}_i - \bar{\mathbf{x}})^T , \tag{5}$$

where n is the number of records in the data set and $\bar{\mathbf{x}}$ is the centroid of the data set. By means of dividing the SSE by the SST we obtain a normalised measure ranged between 0 and 1 that can be easily compared regardless of the used data set. For each method, 10 executions have been run and their averages are shown in Table 5.

4.1 MDAV vs. V-MDAV

The MDAV microaggregation heuristic is a fixed-size heuristic (*i.e.* all the subsets that it generates have the same size). On the contrary, V-MDAV is able to generate subsets of variable size, thus increasing the intra-group homogeneity.

Our intuition that V-MDAV would improve the results of MDAV as a partitioning method has been confirmed with the obtained results shown in Table 5. It is apparent that V-MDAV clearly outperforms MDAV when the records in the microdata set are naturally clustered (*e.g.* see the results of Table 5 for the clustered data sets). The improvements achieved by V-MDAV are not so extraordinary when the records are not naturally clustered – this is the case of the real microdata sets "Census", "EIA", "Tarragona", and the scattered data sets.

Although the main improvements take place on clustered data sets, using V-MDAV on scattered data sets is also beneficial. As shown in Table 5, V-MDAV (along with the GA) almost always obtains the best results.

4.2 One-Step vs. Two-Steps Partitioning

The correct partitioning of the micro data set is essential to obtain a good result (*i.e.* a low information loss). From the results shown in Table 5, it becomes apparent that the 2-step partitioning generally outperforms the 1-step partitioning. This behaviour is specially remarkable when the variable-size microaggregation heuristic is used.

These results confirm that the one-step partitioning splits more natural subsets than the two-step partitioning. By construction, the two-steps partitioning guarantees that small subsets of cardinality k_1 (*cf.* Section 3) will not be split into different parts during the partitioning process. Consequently, the GA can easily optimise the result.

It can be concluded that the two-step partitioning is clearly superior to the one-step partitioning.

4.3 The Role of Dimension

By means of the experiments described above, we have determined that the most promising approach is a two-step hybrid method based on V-MDAV. In this section we briefly analyse the effect of the dimension (i.e. the number of attributed) on this

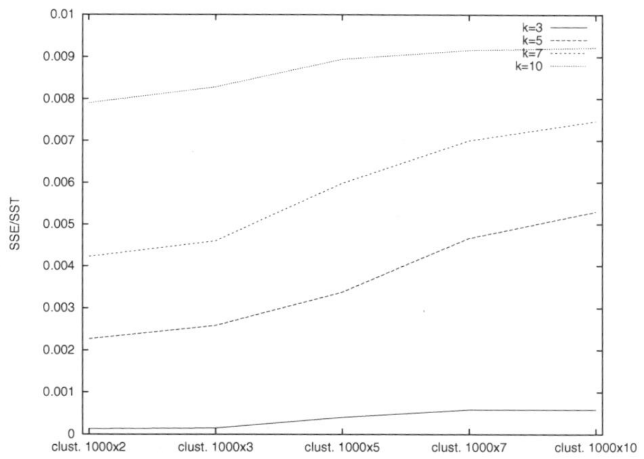

Fig. 5 The effect of the dimension in the performance of the two-step partitioning algorithm with V-MDAV and FUSS over clustered data sets.

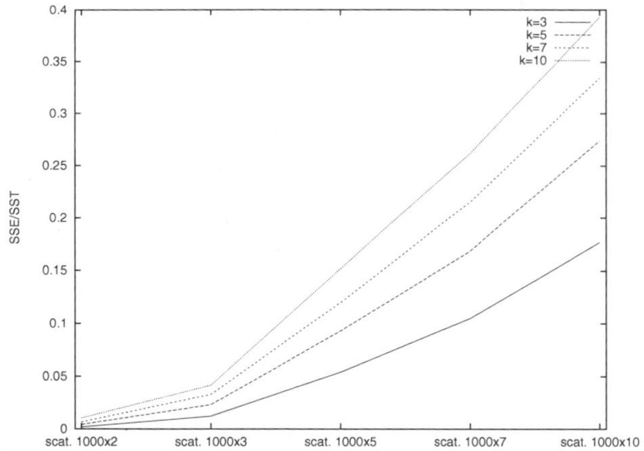

Fig. 6 The effect of the dimension in the performance of the two-step partitioning algorithm with V-MDAV and FUSS over scattered data sets.

specific method for two types of data set (i.e. clustered and scattered) and for several values of k.

To perform these analyses we have generated 5 clustered data sets and 5 scattered data sets with 2, 3, 5, 7 and 10 dimensions each.

Fig. 5 shows the results for the clustered data sets and Fig. 6 for the scattered data sets. It can be observed that an increase in the number of dimensions affects the results obtained by our proposal and produces an increase of information loss. This is specially apparent for scattered data sets.

Also, there are no significant differences, with regard to the growth of the dimension, in the behaviour of the algorithm for the different values of the privacy parameter k.

4.4 The Role of the Size of the Microdata Set

Increasing the number of rows (i.e. individuals in the data set) might lead to the increase of information loss. In this section we analyse the effect of the size of microdata sets on the information loss for clustered and scattered data sets and different values of k. To that end, we have generated 5 clustered and 5 scattered data sets with 1000, 3000, 5000, 7000 and 10.000 elements each.

Figures 7 and 8 show the results for clustered and scattered data sets respectively.

Clustered data sets are barely affected by the increase of the size of the data set. On the contrary, it can be observed that scattered data sets are positively affected by the increase of the size of the microdata set. This behaviour might be due to

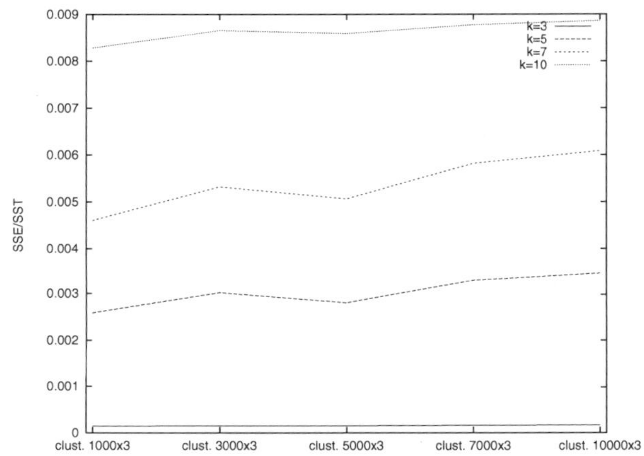

Fig. 7 The effect of the size of data set in the performance of the two-step partitioning algorithm with V-MDAV and FUSS over clustered data sets.

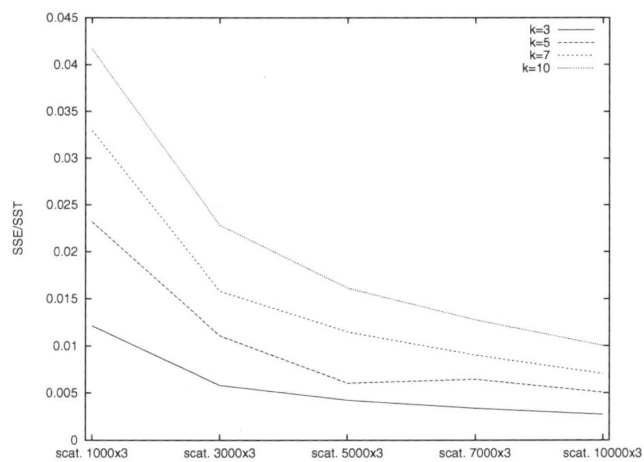

Fig. 8 The effect of the size of data set in the performance of the two-step partitioning algorithm with V-MDAV and FUSS over scattered data sets.

greater growth of the SST with regard to the SSE. However, there is a need for more extensive analyses that are out the scope of this chapter.

5 Conclusions

Due to the NP-hardness of the problem, heuristics are typically used to address the microaggregation of multivariate microdata sets. Microaggregation methods protect the privacy of the respondents by distorting the records in the original data sets whilst, at the same time, they try to reduce the information loss caused by the distortion. In this article, we have shown that using GA to microaggregate data is a promising idea. However, current GA-based methods can hardly deal with large data sets. As a result, mixing classical microaggregation methods to partitioning large data sets with GA seems an interesting solution.

In this chapter we have recalled and analysed some hybrid (i.e. classical microaggregation methods + GA) methods to microaggregate multivariate microdata sets. Following the results of [23] we have shown that the most promising hybrid microaggregation strategy is the two-step V-MDAV mixed with a GA having a selection function based on the fitness uniform selection scheme (FUSS). We have compared MDAV and V-MDAV, and we have concluded that two-steps strategies perform better than one-step strategies. Also, we have briefly analysed the effect of the dimension and the size of the data set on the performance of the microaggregation method proposed.

We plan to study and improve several points in the future, such as:

- Propose methods to find the best values for k_1 and k_2: Similarly to γ, the best values of k_1 and k_2 should vary according to the microdata set. Thus, data-oriented methods seem promising.
- Study new selection strategies that improve the efficiency of the GA and allow the use of larger partitions.
- Extend these methods so as to apply them to non-continuous data.

Acknowledgements. This work was partly supported by the Spanish Government through projects TSI2007-65406-C03-01 "E-AEGIS" and CONSOLIDER INGENIO 2010 CSD2007-00004 "ARES", and by the Government of Catalonia under grant 2009 SGR 1135. The views of the authors with the UNESCO Chair in Data Privacy do not necessarily reflect the position of UNESCO nor commit that organisation.

References

1. Brand, R., Domingo-Ferrer, J., Mateo-Sanz, J.M.: Reference data sets to test and compare sdc methods for protection of numerical microdata. European Project IST-2000-25069 CASC (2002), `http://neon.vb.cbs.nl/casc`
2. Canadian Privacy: Canadian privacy regulations (2005), `http://www.media-awareness.ca/english/issues/privacy/canadian_legislation_privacy.cfm`

3. Defays, D., Anwar, N.: Micro-aggregation: a generic method. In: Proceedings of the 2nd International Symposium on Statistical Confidentiality, Eurostat, Luxemburg, pp. 69–78 (1995)
4. Domingo-Ferrer, J., Martínez-Ballesté, A., Mateo-Sanz, J.M., Sebé, F.: Efficient multivariate data-oriented microaggregation. The VLDB Journal 15(4), 355–369 (2006)
5. Domingo-Ferrer, J., Mateo-Sanz, J.M.: Practical data-oriented microaggregation for statistical disclosure control. IEEE Transactions on Knowledge and Data Engineering 14(1), 189–201 (2002)
6. Domingo-Ferrer, J., Sebé, F., Solanas, A.: A polynomial-time approximation to optimal multivariate microaggregation. Computers & Mathematics with Applications 55(4), 714–732 (2008)
7. Domingo-Ferrer, J., Torra, V.: Ordinal, continuous and heterogeneous k-anonymity through microaggregation. Data Min. Knowl. Discov. 11(2), 195–212 (2005)
8. Domingo-Ferrer, J., Torra, V.: Ordinal, continuous and heterogenerous k-anonymity through microaggregation. Data Mining and Knowledge Discovery 11(2), 195–212 (2005)
9. Edwards, A.W.F., Cavalli-Sforza, L.L.: A method for cluster analysis. Biometrics 21, 362–375 (1965)
10. European Parliament: DIRECTIVE 2002/58/EC of the European Parliament and Council of concerning the processing of personal data and the protection of privacy in the electronic communications sector (Directive on privacy and electronic communications) (July 12, 2002), `http://europa.eu.int/eur-lex/pri/en/oj/dat/2002/l_201/l_20120020731en00370047.pdf`
11. Fayyoumi, E., Oommen, B.J.: A Fixed Structure Learning Automaton Micro-Aggregation Technique for Secure Statistical Databases. In: Domingo-Ferrer, J., Franconi, L. (eds.) PSD 2006. LNCS, vol. 4302, pp. 114–128. Springer, Heidelberg (2006)
12. Hansen, S.L., Mukherjee, S.: A polynomial algorithm for optimal univariate microaggregation. IEEE Transactions on Knowledge and Data Engineering 15(4), 1043–1044 (2003)
13. Holland, J.: Adaptation in Natural and Artificial Systems. University of Michigan Press (1975)
14. Hundepool, A., de Wetering, A.V., Ramaswamy, R., Franconi, L., Capobianchi, A., DeWolf, P.P., Domingo-Ferrer, J., Torra, V., Brand, R., Giessing, S.: μ-ARGUS version 4.0 Software and User's Manual. Statistics Netherlands, Voorburg NL (2005), `http://neon.vb.cbs.nl/casc`
15. Hutter, M.: Fitness uniform selection to preserve genetic diversity. Tech. Rep. IDSIA-01-01, IDSIA, Manno-Lugano, Switzerland (2001)
16. Laszlo, M., Mukherjee, S.: Minimum spanning tree partitioning algorithm for microaggregation. IEEE Transactions on Knowledge and Data Engineering 17(7), 902–911 (2005)
17. MacQueen, J.B.: Some methods for classification and analysis of multivariate observations. In: Proceedings of the 5th Berkeley Symposium on Mathematical Statistics and Probability, vol. 1, pp. 281–297 (1967)
18. Martínez-Ballesté, A., Solanas, A., Domingo-Ferrer, J., Mateo-Sanz, J.M.: A genetic approach to multivariate microaggregation for database privacy. In: ICDE Workshops, pp. 180–185. IEEE Computer Society Press (2007), `http://dx.doi.org/10.1109/ICDEW.2007.4400989`

19. Oganian, A., Domingo-Ferrer, J.: On the complexity of optimal microaggregation for statistical disclosure control. Statistical Journal of the United Nations Economic Comission for Europe 18(4), 345–354 (2001)
20. Samarati, P.: Protecting respondents' identities in microdata release. IEEE Trans. Knowl. Data Eng. 13(6), 1010–1027 (2001)
21. Sande, G.: Exact and approximate methods for data directed microaggregation in one or more dimensions. International Journal of Uncertainty, Fuzziness and Knowledge-Based Systems 10(5), 459–476 (2002)
22. Solanas, A.: Privacy Protection with Genetic Algorithms. In: Success in Evolutionary Computation. SCI, pp. 215–237. Springer, Heidelberg (2008)
23. Solanas, A., Gonzalez-Nicolaas, U., Martinez-Balleste, A.: A variable-mdav-based partitioning strategy to continuous multivariate microaggregation with genetic algorithms. In: The 2010 International Joint Conference on Neural Networks (IJCNN), pp. 1–7 (2010), doi:10.1109/IJCNN.2010.5596660
24. Solanas, A., Martínez-Ballesté, A.: V-MDAV: Variable group size multivariate microaggregation. In: COMPSTAT 2006, Rome, pp. 917–925 (2006)
25. Solanas, A., Martínez-Ballesté, A., Mateo-Sanz, J.M., Domingo-Ferrer, J.: Multivariate microaggregation based on genetic algorithms. In: 3rd IEEE Conference On Intelligent Systems, pp. 65–70. IEEE Computer Society Press, Westminster (2006)
26. Torra, V.: Microaggregation for categorical variables: A median based approach. In: Privacy in Statistical Databases, pp. 162–174 (2004)
27. US Privacy: regulations (2005), `http://www.media-awareness.ca/english/issues/privacy/us_legislation_privacy.cfm`
28. Ward, J.H.: Hierarchical grouping to optimize an objective function. Journal of the American Statistical Association 58, 236–244 (1963)

Breaching Privacy Using Data Mining: Removing Noise from Perturbed Data

Michal Sramka

Abstract. Data perturbation is a sanitization method that helps restrict the disclosure of sensitive information from published data. We present an attack on the privacy of the published data that has been sanitized using data perturbation. The attack employs data mining and fusion to remove some noise from the perturbed sensitive values. Our attack is practical – it can be launched by non-expert adversaries having no background knowledge about the perturbed data and no data mining expertise. Moreover, our attack model also allows to consider informed and expert adversaries having background knowledge and/or expertise in data mining and fusion. Extensive experiments were performed on four databases derived from UCI's Adult and IPUMS census-based data sets sanitized with noise addition that satisfies ε-differential privacy. The experimental results confirm that our attack presents a significant privacy risk to published perturbed data because the majority of the noise can be effectively removed. The results show that a naive adversary is able to remove around 90% of the noise added during perturbation using general-purpose data miners from the Weka software package, and an informed expert adversary is able to remove 91%–99.93% of the added noise. Interestingly, the higher the aimed privacy, the higher the percentage of noise can be removed. This suggests that adding more noise does not always increase the real privacy.

1 Introduction

Data perturbation is a well-studied sanitization method that simultaneously allows access to the data by publishing it and at the same time preserves privacy of the data.

Michal Sramka
UNESCO Chair in Data Privacy, Department of Computer Engineering and Maths, Universitat Rovira i Virgili, Av. Paisos Catalans 26, 43007 Tarragona, Spain and Department of Applied Informatics, Faculty of Electrical Engineering and Information Technology, Slovak University of Technology, Ilkovicova 3, 81219 Bratislava, Slovakia
e-mail: michal@sramka.net

D.A. Elizondo et al. (Eds.): Computational Intelligence for Privacy and Security, SCI 394, pp. 135–157.
springerlink.com

Typically it is achieved using *noise addition*, a technique that adds noise to the data values or query answers which are then released. It can be also achieved through different techniques, but we concentrate on noise addition. Although extensive research has been done in the area of data perturbation, and especially in the context of privacy in statistical databases and statistical disclosure control, released data is still prone to sensitive data disclosures and other privacy breaches.

The additive noise used to perturb data is typically chosen from some distribution. Uniform, normal, exponential and other distributions were considered in the literature. Initially, it was unclear how much noise needs to be added to achieve privacy, and so many ad-hoc approaches were proposed. A systematic treatment came with the introduction of the *differential privacy* notion, which limits the amount of exponential noise that needs to be added to the original data values while providing strong theoretical privacy guarantees.

Data mining is a method to extract implicit, previously unknown, and potentially useful information from data. It has long been used in the area of data privacy, for example, for measuring the usefulness of the released data or more recently for comparing the trade-off between privacy and data utility of different sanitization methods. We have also used it in the past for privacy attacks against anonymized data, in particular for predicting values that may lead to privacy breaches. Likewise, data mining can be used to remove some or all of the added noise from the perturbed released data. Here we present a study of this concept.

We explore the idea of breaching privacy using data mining. Specifically, we use data mining to partially remove noise from perturbed data that satisfies ε-differential privacy. We concentrate on establishing that it is possible to remove the majority of the added noise using data mining techniques alone, without using or assuming any external knowledge (in literature sometimes also referred to as background, auxiliary, side, or external knowledge or information). We refer to such attacks and the corresponding adversaries as *naive*. The naive adversaries thus represent individuals without external knowledge or expertise in data mining technology, and can be seen as a variation of the "script kiddies" that use prepared scripts and tools without really understanding the ideas or the technology behind them. Our adversarial model captures both naive adversaries as well as much stronger adversaries – adversaries that possess external knowledge about the published data and/or adversaries having expertise in the field of data mining. Indeed, data mining experts would not use general purpose tools, but rather they would first analyze the data and then use specific data mining algorithms with fine tuned parameters. Capturing and establishing bounds for the amount of external knowledge and expertise is a well-known problem in data privacy. Without the need to limit these unknowns, we demonstrate that a simple collection of external data together with data mining and fusion expertise is enough to remove more than 99% of the noise that was added to the sensitive values and that was supposed to protect the privacy of the concerned individuals.

We illustrate the scenario and the various adversaries on the following example: A data owner wants to release its data collection about individuals while protecting the privacy of their ages. To achieve the privacy, the data owner sanitizes the data by adding random noise (e.g., limited by the ε-differential privacy) to all the ages individually, and releases the sanitized data. An adversary launches the attack by applying one or more data mining algorithms to uncover hidden trends and patterns in the data. In the simplest case, a naive adversary simply applies some data mining algorithms directly to the sanitized data and fuses the outputs. An expert adversary would in addition select particular data mining algorithms that best fit the data at hand. The mining results allow the adversaries to make predictions of ages. An informed adversary would obtain similar data of unrelated individuals, apply data mining to it and use this mining results to make predictions of ages in the sanitized data. In all these cases, the predictions of ages should be seen as the estimates of the original ages and thus the added noise can be partially removed from the sanitized data. This estimation and partial noise removal results in the breach of privacy.

1.1 Our Contribution

We propose a comprehensive adversarial model and a noise removal process based on data mining and fusion methods. Basically, multiple data mining algorithms are applied to the released sanitized data or external data and the mined knowledge is used to get multiple predictions of each of the sensitive values in the sanitized data (e.g., age). These multiple predictions are fused together and the combined value then represents an estimate of the original sensitive value. An estimate that is possibly closer to the original value, in other words, a value that may have some of the noise removed.

The proposed attack relies on data mining [32] and fusion techniques [30, 5]. Depending on the skills, knowledge, and information an adversary possesses, our model encompasses the following three basic types of adversaries: naive, informed, expert, and one combination: informed+expert adversary. The goal of each adversary is to breach privacy of some individual(s) in the sanitized data.

- A *naive adversary* does not have any external knowledge about the sanitized data and does not have any expertise in the data mining field. The naive adversary can be seen as a variation of the "script kiddies" that use prepared scripts and tools without really understanding the ideas and technologies behind them. However, the attack by a naive adversary is very practical – it can be launched by adversaries that have no other information except the sanitized data and have no or very little knowledge about data mining and fusion, e.g., reporters. Indeed, simulations using some readily-available data mining programs and using simple average for fusing outputs demonstrate that even these non-expert not-knowledgeable adversaries and their simple techniques remove remarkable amount of noise and breach privacy.
- An *informed adversary* possesses external data that can be used for data mining and successively the mined knowledge is used for prediction of the perturbed values of the sanitized data. Such external data can be collected for individuals not related to the individuals about whom the sanitized data is. The purpose of collecting or otherwise obtaining the external data is purely to help breach

privacy of the individuals in the sanitized data, not to obtain data that can be used for the same or similar purpose as the original data. We recognize that the option to collect or obtain external data is not always available to the adversary. However, the possibilities for assuming possession of external data by an adversary are limitless in theory. In our experiments, we assume simple external knowledge and show that it is often enough to remove the vast majority of the noise added during perturbation.
- An *expert adversary* has expertise in data mining and fusion technologies. Such adversary is assumed to use specific and tailored data mining and fusion techniques that best fit the structure, amount, and type of the data at hand. In case an expert adversary also possesses external data, we speak of an *informed+expert adversary*. Again, capturing the expertise of such adversaries is a problem, because the possibilities for assumptions are limitless in theory. We consider the unbounded expert adversary in our model, but in the experiments we assume a concrete expert adversary having just common (limited) knowledge in the field of data mining.

We focus on proposing a general adversarial model that encompasses all these types of adversaries. The model assumes that privacy is achieved by additive noise perturbation of individual sensitive values. We limit the amount of additive noise being added to the original values during perturbation by employing the ε-differential privacy, but our general model is independent from this privacy notion.

Our extensive experiments are performed on four different data sets derived from the UCI Machine Learning Repository's Adult and IPUMS97 census-based databases [4]. The ages of the individuals are perturbed using ε-differential privacy [9, 12] that adds noise chosen from a Laplace distribution. We simulate the adversary by applying up to six data mining algorithms from the Weka software package [32] to the sanitized or similar unsanitized data and fuse the outputs using different fusion strategies. We evaluate the success of the proposed attack by measuring the amount of noise that is removed by considering the fused values as age estimates, and the "nearness" of these estimates to the original values of age. We require a high privacy from the perturbation, namely $\varepsilon \in \{0.5, 0.1, 0.05, 0.01\}$. The similar leakages between 0.0001 to 1 have been used by one of the differential privacy co-inventors in [19], although some other illustrations in literature only use weaker privacy with leakage $\varepsilon = 2$ [1] or even overall procedure leakage $\varepsilon = 8.6$ [18]

The naive adversaries are the most simple, non-expert adversaries without external knowledge that use general-purpose data mining techniques and a simple fusion that computes the average. Yet, the experiments show that such naive adversary removes 82%–94% of the added noise in the attack. The risk and consequences of this attack is substantial because no external knowledge is used in it, and thus any external knowledge can only aid the adversary. This scenario is simulated next by an informed+expert adversary that according to our experimental results removes 91.05%–99.93% of the noised added during perturbation. An important observation we make from the experiments is that the higher the aimed privacy of the perturbation/sanitization, the higher percentage of the noise can be removed. In other words, adding more noise does not necessarily increase privacy in these scenarios. Overall, the experimental results confirm that our attack with any type of adversarial knowledge and expertise presents a significant risk to statistical disclosure and privacy.

We note that we do not understand the experimental results as a fallacy of the theoretical guarantees of differential privacy. The differential privacy notion is an important and a strong theoretical privacy definition. It is based on probabilistic concepts. As with every probabilistic and statistical tools, one has to be careful about how are the tools applied. We consider scenarios with specific query functions that may be typically chosen by data owners to protect privacy. We show that in these scenarios it is possible to remove the majority of the added noise from perturbed released data. Our attacks are applicable to ε-differentially perturbed data as well as to data sanitized with other perturbation techniques. Finally, we briefly discuss some countermeasures that should be considered in real-world applications to avoid these kinds of attacks.

1.2 Organization of the Chapter

We finish the Introduction with overview of related work in the literature. Next, we briefly outline the concepts of perturbation, noise addition and differential privacy together with basic notation and terminology in Sect. 2. Our general adversarial model is presented in-depth in Sect. 3. It starts by extending the example from the Introduction, continues with the description and discussion of the attack and various levels of adversaries, and finishes with evaluation criteria for the attack. Sect. 4 contains comprehensive experimental evaluation on four census databases and assuming various strength of adversaries. We sum up and conclude in Sect. 5, where we also discuss some basic countermeasures against these attacks and a few open research topics.

1.3 Related Work

Data perturbation and noise addition has been identified early for achieving data privacy [7, 2]. Uniform and normal (Gaussian) distributions are used in [3] to sample the additive noise, while exponential (Laplace) distribution is used to achieve differential privacy [9, 12]. Differential privacy is a well-studied notion [6, 17, 10, 11, 13, 20], which provides strong theoretical guarantees about the privacy of the released data. Differential privacy is usually achieved by noise addition, however a recent research claims that noise addition is not enough [21, 22, 23]. We use the differential privacy notion to limit the amount of (exponential) noise that needs to be added to original values in order to achieve measurable levels of individual's privacy.

Data mining has been traditionally considered in the area of data privacy for measuring the usefulness of the released data [3, 16, 28]. It has also been used to measure not just the usefulness of the released data but also the amount of privacy various sanitization methods provide [27, 25]. Data mining has also been used to attack the privacy of the data [31, 26]. Similarly, we use data mining to attack the privacy by removing portions of the added noise. In other words, we use data mining not to evaluate the usefulness of the released sanitized data, but to attack the privacy of the data. In the preceding conference paper [24] we concentrated on establishing that it is possible to remove the majority of the added noise using data mining techniques alone on the sanitized data, without using or assuming any external information. Here we extend our adversarial model to encompass stronger adversaries and capture their characteristics and abilities. Namely, we focus on informed adversaries having external data that can be used for predictions over sanitized data, expert adversaries having knowledge and experience in the fields of data mining and fusion, and the most powerful adversary – the combination of the informed and expert adversaries.

It is possible to statistically estimate original values [7, 2] when repeating queries and obtaining different noisy values. In contrast, if the noise is fixed for the same value [3], an adversary is unable to improve the estimate using just the simple statistical methods – we propose an attack to estimate the original values in this case using data mining.

Similar to our attack that removes noise by estimating the original values, [8] presents an attack that obtains disclosures by approximating the original data values from matrices distorted by addition of random values. The authors use matrices and a specific matrix-based spectral filtering techniques to obtain disclosures, while we use relational databases and simple general-purpose data mining techniques. A web-based information-fusion privacy attack was mentioned in [14]. It considers the fusion of released k-anonymized data with external information obtained from the web (possibly including "human-in-the-loop"), but the paper lacks any details or experimental results showing the effectiveness or impact of the proposal.

Combining data from several measurements is referred to in the literature as data fusion, data aggregation, data consolidation, and data integration. We only use a simple fusion strategy to combine outputs of multiple data miners, but there is a lot of research in the field of fusion, aggregation of data mining results, and verification methods [30, 5, 29, 15].

2 Differential Privacy

A *database DB* is a set of tuples, where a *tuple*

$$x := (x_1, \ldots, x_t)$$

consists of *fields* x_i's defined over some finite domains (categorical data) or infinite domains (numerical data). This concept conforms to the traditional relational database definition. We refer to the database *DB* as the *original data* and assume it is in possession and control of a *data owner*.

A *sanitization mechanism* S is a randomized algorithm that for a database DB and query f returns $\mathsf{S}(f(DB))$, a transformation of the query result $f(DB)$ that conforms to the "privacy guarantee" of S. A *privacy guarantee* of S is a notion, a description of what is meant and provided by the sanitization mechanism and how sensitive data, identities, and sensitive relations are protected by S.

For simplicity, from now on, we assume that the data owner uses a sanitization mechanism S to release the original data DB without suppressing any tuples. We further suppose, although this time without loss of generality, that suppressed fields are replaced with dummy values. Hence, there is a one-to-one correspondence between the fields and tuples of DB and their sanitized versions. Then by $DB^\star$, $x^\star$, and $x_i^\star$, respectively, we denote the database DB, tuple x, and field x_i after sanitization, see Table 1. We refer to $DB^\star$ as the *sanitized data* or *perturbed data*, and we assume this data is public – available to *data users* that can be legitimate (analysts) or malicious (adversaries).

Table 1 Notation: The sanitization mechanism S maps DB to $DB^\star$ and preserves one-to-one correspondence between the fields and tuples

original		**sanitized**
data DB	$DB \overset{\mathsf{S}}{\mapsto} DB^\star$	(perturbed) data
tuple $x \in DB$	$x \mapsto x^\star$	tuple of $DB^\star$
field value x_i of tuple x	$x_i \mapsto x_i^\star$	value of tuple $x^\star$

The *differential privacy* is a notion based on the indistinguishability of transcripts (messages and outputs of a data user interacting with a sanitization mechanism). In the database setting, it is typically achieved by adding random Laplace noise to numeric values, and it effectively limits the amount of necessary noise needed for achieving privacy. The privacy guarantee of the differential privacy notion is semantic, that is, it considers the shifts and incremental gain in the adversary's knowledge.

Definition 1 (ε-differential privacy [9]). A randomized function S gives *ε-differential privacy*, if for all databases DB_1 and DB_2 differing on at most one element and all $R \subseteq Range(\mathsf{S})$ holds that

$$\Pr[\mathsf{S}(DB_1) \in R] \leq \exp(\varepsilon) \times \Pr[\mathsf{S}(DB_2) \in R] \ .$$

The ε-differential privacy notion limits the probability that the randomized function (sanitization mechanism) S would leak information from a database that is differing by at most one element. Therefore a data leak and a disclosure of a private data through S is possible, but the probability of such leak is limited through the *leakage* parameter ε.

In practice, the sanitization mechanism achieving ε-differential privacy for numerical values works by adding noise to the query result $f(DB)$, a noise that is randomly chosen from the Laplace (exponential) distribution $\mathrm{Lap}(0, \Delta f/\varepsilon)$ with the probability density function

$$h(z) = \frac{\varepsilon}{2}\exp(-\frac{\varepsilon|z|}{\Delta f}) \ ,$$

where Δf is the sensitivity [9] of the query function f defined as

$$\Delta f = \max_{DB_1, DB_2} \|f(DB_1) - f(DB_2)\|_1$$

with DB_1 and DB_2 again ranging over all databases differing on at most one element.

If fresh noise is sampled and added to each query result, then it is possible to statistically estimate the original values [7, 2] after repeating the same queries. Therefore, it has been suggested (e.g., in [2, 3]) to fix the noise for the already seen values and only add freshly sampled noise for the new, previously unseen values. Similarly, our noise-adding function "noise", for a fixed query f and a fixed leakage ε, depends only on the original value, and returns the same noise for already seen input values and for new input values it returns a fresh sample from the Laplace distribution $\mathrm{Lap}(0, \Delta f/\varepsilon)$. If v_i is the original sensitive value (field value of a tuple, for some i) that needs to be protected and $v_i^\star$ denotes the perturbed protected value, then

$$v_i^\star := v_i + \mathrm{noise}(v_i) \ .$$

This thwarts the statistical attack, as it does not allow an adversary to improve the estimates by repeating queries. We also focus on this scenario and show that data mining can improve the estimates of the original values.

3 The Attack Model

To make the explanation more concrete, we start by presenting our attack model on an example that uses ages. However, the attack would work for any numerical attribute.

A data owner wants to release its data collection about individuals. The data owner considers the individuals' ages to be a sensitive information. The owner protects the privacy of the ages by adding random noise limited by the ε-differential privacy to all the ages individually, and then releases the sanitized data. Formally, suppose that the original data is DB, and that noise is added during sanitization to x_m, the m-th field of every tuple $x \in DB$ that represents age. The added noise is different for distinct ages, but fixed for the same values of age. The newly formed sanitized tuples $x^\star$'s are then released as the sanitized data $DB^\star$. The sanitized data can then be accessed by the data users – legitimate data users (analysts) as well as malicious data users (adversaries).

An *adversary* is an end user of the sanitized data with malicious intentions, namely trying to make disclosures and breach privacy. An adversary launches the attack by applying one or more data mining algorithms to uncover hidden trends and patterns in the data available to the adversary that would allow him/her to partially remove the added noise and obtain quality estimates of the ages.

The attack consists of first obtaining predictions for the ages that have been perturbed and that are of the interest to the adversary. Data miners can provide such predictions. However, multiple data miners produce multiple predictions. Moreover, there may be different predictions, based on individual records in the database, for the same original value of age. Therefore the adversary combines the different predictions obtained from multiple miners using fusion methods. The adversary's intention is to use all these predictions and combinations to obtain reasonable estimates of the ages – estimates that contain much less noise than the perturbed values and possibly estimates that breach privacy.

A *naive adversary* launches the attack with readily-available (random selection of) data miners that mine over the sanitized data and combines the different predictions with simple average computation. Of course, building knowledge for prediction just from the sanitized data may not exceed the sanitized data itself. Therefore we also consider a stronger, *informed adversary* who possesses external data over which the knowledge for prediction is mined and then applied to the sanitized perturbed values. Also, selecting the data miners at random and using average for fusion is suboptimal. In the field of data mining, specific approaches and techniques exist that are beneficial for different data sets. An *expert adversary* has the knowledge and expertise in data mining and fusion (aggregation of data mining results). Finally, the most powerful adversary is the *informed+expert adversary*, the combination of the last two adversaries.

3.1 Noise Removal Using Data Mining and Fusion

A data miner or simply a *miner* M is an algorithm that takes either the sanitized data $DB^\star$ or an external data DB^{ext} (owned by an informed adversary) as the input together with one sanitized tuple $x^\star$. The miner determines trends and patterns in this input data and outputs a prediction for the m-th field $x^\star_m$ of the tuple

$$x^\star = (x^\star_1, \ldots, x^\star_m, \ldots, x^\star_t) \ .$$

The data structure of $x^\star$ is the same as of $DB^\star$, but not necessarily the same as of DB^{ext}. The mining process described by M should be understood as obtaining knowledge (learning) from $DB^\star$ or DB^{ext} and then applying the learned knowledge to $x^\star$ to make predictions about $x^\star_m$. In practice this means that some of the fields (attributes) of DB^{ext} should be the same as in $DB^\star$. Prediction of $x^\star_m$ is then based precisely on these fields.

An informed adversary has or obtains DB^{ext}. The purpose of DB^{ext} is not to replace the utility of DB or $DB^\star$, but purely to make disclosures and breach privacy of individuals in DB. The amount of external knowledge in DB^{ext} can be limitless in theory. A theoretical result states [9] that there is always some external knowledge that by itself is not a disclosure but by combining it with the sanitized data it becomes a disclosure. Therefore it makes sense to limit the amount of external knowledge available to the informed adversary. We do not limit the external knowledge explicitly in our model, but in experimental evaluation we demonstrate that an informed adversary possessing similar data to *DB* can successfully remove substantial amount of noise from perturbed values.

The adversary possesses a set of n miners $M = \{\mathsf{M}_1, \ldots, \mathsf{M}_n\}$, each predicting the same field m of any input tuple. We consider m to be an implicit parameter of each miner M_i. The *prediction* for the m-th field of the tuple $x^\star$ obtained by the miner M_i is denoted by p_i, that is, for a naive adversary

$$p_i \leftarrow \mathsf{M}_i(DB^\star, x^\star) \enspace ,$$

and for an informed adversary

$$p_i \leftarrow \mathsf{M}_i(DB^{\text{ext}}, x^\star) \enspace .$$

After the adversary performs all the mining, the adversary will obtain n predictions for the m-th field for each tuple $x^\star \in DB^\star$.

The miners M_i's can be any data mining algorithms. A naive adversary would select a few general-purpose miners that are freely available and are easy to use. In practice, this selection suffices to launch a successful attack. An expert adversary would, however, select, modify and parametrize miners specifically for the type, structure and amount of data at hand (either $DB^\star$ or DB^{ext}).

At this step of the attack, when considering any single miner M_i, the adversary (naive, informed, or expert) possesses a prediction p_i for the field m for every tuple $x^\star \in DB^\star$. A single original value x_m, that is by definition perturbed into

$$x_m^\star = x_m + \text{noise}(x_m) \enspace ,$$

may appear in multiple tuples in the database *DB*. For example, think about age – there are about 120 ages that are repeated for individuals in the database. Consequently, because of the properties of the noise function, the same perturbed value $x_m^\star$ appears in multiple tuples in the database $DB^\star$. Therefore predictions for the same original value x_m may be different over different tuples $x^\star$ having the same value $x_m^\star$, because the predictions are based not just on the filed $x_m^\star$ but on multiple or all the fields of the tuple $x^\star$. Thus the adversary possesses different predictions for each single $x_m^\star$. The adversary proceeds with the attack by first fusing these different predictions for each distinct value of $x_m^\star$ and then the adversary fuses the values from different miners.

Let Y be the set of all distinct values x_m from all the tuples in *DB*. Then clearly

$$Y^\star := \{x_m^\star \mid x_m^\star = x_m + \text{noise}(x_m) \ : \ x_m \in Y\}$$

is the set of all distinct perturbed values. Because of the randomness of the noise selection in practice, we can assume that no two distinct original values are perturbed into the same perturbed value, which is the same as assuming that $|Y| = |Y^\star|$.

A *fusion* is an algorithm that takes multiple inputs, fuses them, and outputs a single combined value. The adversary uses a fusion algorithm F_1 to combine the different predictions by a single miner that were obtained for the same value of $x^\star_m$. The adversary does this for all distinct perturbed values and for all miners. Algorithmically, for each miner $\mathsf{M}_i \in M$, and for each each distinct perturbed value $y \in Y^\star$, the adversary computes

$$q_{i,y} \leftarrow \mathsf{F}_1(\{\, p_i \mid p_i = \mathsf{M}_i(DB^\star, x^\star) \ : x^\star \in DB^\star \text{ having } x^\star_m = y \,\}) \ .$$

This computation results in n predictions (one for each miner) for each distinct perturbed value $x^\star_m$.

Finally, the adversary fuses the multiple predictions from multiple miners into a single prediction r_y for each perturbed value $y \in Y^\star$ using a fusion algorithm F_2 as follows:

$$r_y \leftarrow \mathsf{F}_2(\{\, q_{i,y} \mid i = 1, \ldots, n \,\}) = \mathsf{F}_2(q_{1,y}, \ldots, q_{n,y}) \ .$$

The fused value r_y represents an *estimate* of the perturbed value y.

The adversary ends with a database derived from $DB^\star$, where $x^\star_m$ of each tuple is replaced with its estimation $r_{x^\star_m}$. This concludes the attack.

The fusion algorithms F_1 and F_2 can be based on any fusion strategy [30, 5, 29, 15]. A naive adversary uses a simple average for fusions, while an expert adversary may use advanced fusion/aggregation techniques, such as weighted average, voting, stacking, bagging, etc. In our experiments, we use simple average and weighted average (for an expert adversary) with remarkable success.

3.2 Evaluation of the Attack

The adversary is clueless with regard to which predictions and estimates are correct, if any at all. We evaluate the success of the attack from the privacy viewpoint, that is, if there are privacy breaches and disclosures. This evaluation covers the impact the attack has for the data owner trying to protect the privacy of the data as well as for the adversary who is interested to remove the noise from the perturbed data.

There are several possibilities to evaluate the success of the attack. We focus on how much of the privacy protection provided by the data perturbation is lost due to the attack. We have previously developed a comprehensive way to measure success of the data mining methods over sanitized/perturbed data based on the data mining utility [27, 26]. The previously proposed utility functions could measure the success of predictions, and are flexible enough to measure the success of this attack. But contrary to our needs in this attack, the previously proposed utility functions measure and evaluate the success of predictions per each tuple (per each record) in the database. Here we concentrate on measuring of how much noise is being removed by predictions from the perturbed values that may be the same for many tuples. We borrow the "nearness" concept from [27, 26] and also evaluate the success by looking at the amounts of added and remaining noise.

First, we focus on the success of the estimation of the distinct values $x^\star_m$'s. An estimation $r_{x^\star_m}$ is *δ-nearer* with respect to the original value x_m, if

$$d(r_{x^\star_m}, x_m) \leq (1-\delta)\, d(x^\star_m, x_m) \ ,$$

where d is a distance function, for example, the Euclidean distance function. We say an estimation $r_{x^\star_m}$ is *nearer* w.r.t. x_m, if it is δ-nearer for any $\delta > 0$, which translates into strict inequality:

$$d(r_{x^\star_m}, x_m) < d(x^\star_m, x_m) \ .$$

In other words, the "nearness" tells us whether the estimation $r_{x^\star_m}$ is a better representation for x_m than the perturbed value $x^\star_m$. For example, age 20 is perturbed into 31.4. Then 25 is a nearer estimate, and so is 18. Age 32 is not a nearer estimate. Age 21 is a 0.9-nearer (or 90%-nearer) estimate, because it lies within the $1/10^{\text{th}}$ of the distance $|31.4 - 20|$ from the original age 20. Our measure of success of the noise-removing attack is the number of estimates that are nearer in comparison to all nearer and non nearer estimates, and we will also look at the δ factor of these estimates.

For our other measure of success we compute the amount of noise that has been added to the different values during perturbation and compare it with the amount of noise that remains in the estimated values. The *cumulative added noise* is

$$\text{added noise} := \sum_{x_m \in Y} |x^\star_m - x_m| \ ,$$

where we decided to sum over all distinct values rather than over all tuples $x \in DB$, and the *cumulative remaining noise* after the attack is

$$\text{remaining noise} := \sum_{x_m \in Y} |r_{x^\star_m} - x_m| \ .$$

Our second measure of the success of the noise-removing attack is the ratio of the remaining noise to the added noise.

Existing verification and validation methods for data mining and fusion techniques [30] can help the adversary to evaluate the success of the attack, even without having access to the original data. Note that our evaluation cannot be computed by the adversary because it uses the original values not available to the adversary. The same techniques can suggest to the adversary which of the data miner is the best for the noise removal. However, we abstracted for the need to identify the best miner, as we fuse the various predictions.

4 Experimental Results

Our experiments demonstrate the power of our adversarial model. The privacy attacks are successful regardless of the choice of a database and regardless of the privacy parameters believed to provide very strong privacy guarantees. Of course, the success increases with the strength of the adversary. A naive adversary mining over sanitized data and averaging the outputs to obtain estimates is able to remove around 82%–94% of the added noise while having 85%-96% "nearer" estimates. An informed adversary having and mining over similar external data and averaging the outputs of predictions over sanitized data removes 89%–99.86% of the noise and obtains 91%–100% "nearer" estimates. If the informed adversary also has some data mining and fusion expertise, then such informed+expert adversary that uses specific miners and fuses predictions based on the weights s/he assigns to the miners removes 91%–99.93% of the noise and gets 92%-100% "nearer" estimates.

We performed the experiments on four databases derived from two census-based UCI's Machine Learning Repository [4] databases, namely the Adult and IPUMS databases. We sanitized the age of the individuals in these databases using data perturbation (noise addition) limited by the ε-differential privacy notion [9]. We used $\varepsilon = 0.5, 0.1, 0.05$, even 0.01, which represents increasingly stronger privacy, similar to [19] and much stronger than used in some other experiments [1, 18]. We simulated the adversary by using up to six miners from the Weka software package [32] and by using simple and weighted average as the fusion strategies to combine multiple values. This comprehensive simulation of attacks demonstrate that even simple techniques (average) and readily-available tools (Weka miners) can remove the majority of the noise that have been added to individuals' ages during perturbation, and that assuming some expertise in data mining and possession of some external data, almost all of the noise can be removed. Interestingly, the smaller the leakage ε (the higher the aimed privacy), the higher the percentage of the removed noise independent of the strength of the adversary.

4.1 Data and Sanitization

We derived four different databases from the UCI Machine Learning Repository's Adult and IPUMS97 [4] census-based data:

1. *Adult-UCI* consisting of 45,222 records and 15 attributes was the combination of learning and testing data of the Adult data set, excluding the records containing missing values.
2. *Adult-Iyengar* was derived from *Adult-UCI* by removing some attributes as proposed in [16]. This data having 9 attributes is a "benchmark" for experimental testing in the privacy field.
3. *IPUMS97-num* consisting of 70,187 records each having 61 attributes was a sample of Los Angeles and Long Beach area citizens from 1997. All 61 attributes were considered to be numerical.

4. *IPUMS97-cat* was the same as *IPUMS97-num*, except all the attributes but age were considered to be categorical.

The databases contained census-based information about individuals. In particular, individuals are described using several attributes that differ over the databases, but in all cases the attributes include *age*. All the data and attributes are described in detail in [4] and [16], respectively.

We split each databases into two parts DB and DB^{ext} having roughly the same number of records. This way we could simulate the sanitization and possession of external data by adversaries at the same time and uniformly for all types of adversaries.

To obtain the perturbed sanitized data, we used ε-differential privacy in order to limit the noise that was being added to the attribute *age* of each of the four databases. We used leakage $\varepsilon = 0.5$, 0.1, 0.05, and 0.01, which all represent a strong privacy protection. Then the additive noise was chosen randomly from the Laplace distribution $\mathrm{Lap}(0, \Delta f/\varepsilon)$ with the probability density function

$$h(z) = \frac{\varepsilon}{2}\exp(-\frac{\varepsilon|z|}{\Delta f}) \ ,$$

where Δf is the sensitivity [9] of the query function f, which basically describes how much can query results differ on two queries. The query function f in our case was a function which "selects" age, so $\Delta f = 73$ for the two Adult databases, where the ages ranged between $17 - 90$, and $\Delta f = 100$ for the two IPUMS97 databases, where the ages ranged between $0 - 100$. The noise was fixed for each age, so that observations of multiple records cannot improve an adversary's estimate of the age [2, 3]. Using the notation developed in the previous sections, the original age in any database tuple x is denoted as x_{age}, and the perturbed age in a sanitized tuple $x^\star$ is denoted as $x^\star_{age}$.

4.2 Attacks

The following six miners representing three different types of miners were selected from the Weka [32] software package, version 3.6.2, and used in the attacks by the naive and informed adversaries:

- function-based: RBFNetwork (M_1)
- rule-based: ConjunctiveRule (M_2), DecisionTable (M_3), and ZeroR (M_4)
- tree-based: DecisionStump (M_5) and REPTree (M_6)

As mentioned earlier, these are general-purpose miners. The Weka user interface allows adversaries to use them with no or minimum knowledge about data mining. On the contrary, the informed+expert adversary that we simulated in these experiments used only three specific miners and ranked them with a weight:

- the rule-based DecisionTable (M_3) with weight $w_3 = 2$,
- the tree-based DecisionStump (M_5) with weight $w_5 = 1$, and
- the tree-based REPTree (M_6) with weight $w_6 = 4$.

Table 2 Types of adversaries, their data and tools

Task \ Adversary:	Naive	Informed	Informed+expert
Miner set M:	$\mathsf{M}_1,\ldots,\mathsf{M}_6$	$\mathsf{M}_1,\ldots,\mathsf{M}_6$	$\mathsf{M}_3,\mathsf{M}_5,\mathsf{M}_6$
Mining over:	$DB^\star$	DB^{ext}	DB^{ext}
Fusion F_1:	average	average	average
Fusion F_2:	average	average	weighted average

The higher the weight the more precise predictions were expected by the adversary for these types of data. These weights associated with these miners are a result of the knowledge and expertise of the adversary in data mining. The expert adversary is assumed to know how the miners perform over the data at hand (size of database, number and type of fields, range of values, etc.) and so can rank the miners appropriately.

The adversaries applied the miners to different data. The naive adversary only had the sanitized data $DB^\star$, so applied the six miner to it. The informed and informed+expert adversaries possessed external data, here simulated by the set DB^{ext}, and applied their miners to this external data. Table 2 tracks the differences among the adversaries.

Because fresh noise was added only to new values and fixed for the same values, an adversary could identify the records $x^\star$ in the perturbed released data that have the same age $x^\star_{age}$ and compute the average of the corresponding predictions. Note that although the added noise was same for the same age values, the data mining predictions may have varied even for the same age, because the predictions were based on the whole tuple $x^\star$.

Using the notation from Section 3.2, the set Y consisted of all ages in the database and the set $Y^\star$ of all the perturbed ages: $|Y| = |Y^\star| = 74$ for the two Adult databases and $|Y| = |Y^\star| = 101$ for the the IPUMS97 databases. The fusion algorithm F_1 simply took all the possibly various predictions p_i's by one miner M_i that were obtained by this miner over all those tuples $x^\star$ that have the same perturbed age $y := x^\star_{age}$ and computed the average. Formally, $q_{i,y}$ is computed as follows:

$$q_{i,y} := \sum \frac{p_i}{c} \quad ,$$

where the sum is over all those p_i's such that

$$p_i := \mathsf{M}_i(DB^\star, x^\star) \quad \text{or} \quad p_i := \mathsf{M}_i(DB^{\text{ext}}, x^\star) \quad ,$$

for naive and informed adversaries, respectively, computed for all those $x^\star$ that have the age attribute value equal to y ($x^\star_{age} = y$) and where c is the number of such $x^\star$'s. This was repeatedly computed for all distinct perturbed ages y from among $x^\star_{age}$'s, and for all miners M_i. For each distinct perturbed age $y = x^\star_{age} \in Y^\star$, this resulted in six predictions $q_{1,y}, \ldots, q_{6,y}$ in the case of naive and informed adversaries, and in three predictions $q_{3,y}, q_{5,y}, q_{6,y}$ in the case of the informed+expert adversary.

The second fusion algorithm F_2 for the naive and informed adversaries took the six combined predictions obtained as a result of mining with six miners, and fused them once more, again using simple average:

$$r_y := \frac{q_{1,y} + \cdots + q_{6,y}}{6} \quad .$$

In the case of the expert adversary, the second fusion algorithm F_2 was given by weighted average according to the weights that represent the trust of the adversary in the outputs from the miners:

$$r_y := \frac{w_3 q_{3,y} + w_5 q_{5,y} + w_6 q_{6,y}}{w_3 + w_5 + w_6} = \frac{2q_{3,y} + q_{5,y} + 4q_{6,y}}{7} \quad .$$

Table 2 summarizes the differences among the adversaries. In each case though, the average $r_{x^\star_{age}}$ represented a fusion of multiple predictions, and we considered it to be an estimate of the original age x_{age}.

Regarding the time complexity and computational cost of the attack, the mining dominated over all other tasks. Since we used a simple fusion, it only took negligible time (a few seconds). On quad Intel Xeon 3.0 GHz, the whole prediction consisting of training and testing the selected six data miners took on average 30 seconds, 50 seconds, 6.8 minutes, and 7.5 minutes for *Adult-Iyengar*, *Adult-UCI*, *IPUMS97-num*, and *IPUMS97-cat*, respectively. This time increase is due to the number of fields (and their types) and the number of tuples of the respective databases. The average times were computed for different splits of the data into DB and DB^{ext} and for all sanitizations using given leakages ε.

4.3 Evaluation and Discussion

Nearer estimates comprise a privacy breach. We considered an estimate $r_{x^\star_{age}}$ to be nearer with respect to the original age x_{age} if

$$|r_{x^\star_{age}} - x_{age}| < |x^\star_{age} - x_{age}| \quad .$$

Table 3 presents the number of nearer estimates from the attacks. Between 85%–96% out of all estimates by the naive adversary are nearer. Not all estimates are nearer, because probably there was not enough data for some ages in the data and so the data miners could not always produce highly accurate and confident predictions. The informed and informed+expert adversaries were able to get almost all estimates (91%–100%) nearer. Interestingly, the smaller the leakage ε (the higher the aimed privacy), the higher the number of nearer estimates independent of the adversary. Similar situations are observable also in other evaluations.

Table 3 Number and percentage of the "nearer" estimates (out of possible 74 for the Adult and 101 for the IPUMS97 databases) obtained by different adversaries

(a) Naive adversary

DB \ Perturb	$\varepsilon = 0.5$	$\varepsilon = 0.1$	$\varepsilon = 0.05$	$\varepsilon = 0.01$
Adult-UCI	63 (85%)	71 (96%)	71 (96%)	71 (96%)
Adult-Iyengar	64 (86%)	68 (92%)	67 (91%)	67 (91%)
IPUMS97-cat	92 (91%)	95 (94%)	95 (94%)	95 (94%)
IPUMS97-num	90 (89%)	93 (92%)	93 (92%)	93 (92%)

(b) Informed adversary

DB \ Perturb	$\varepsilon = 0.5$	$\varepsilon = 0.1$	$\varepsilon = 0.05$	$\varepsilon = 0.01$
Adult-UCI	67 (91%)	72 (97%)	72 (97%)	72 (97%)
Adult-Iyengar	67 (91%)	72 (97%)	72 (97%)	72 (97%)
IPUMS97-cat	95 (94%)	100 (99%)	100 (99%)	101 (100%)
IPUMS97-num	97 (96%)	100 (99%)	100 (99%)	101 (100%)

(c) Informed+Expert adversary

DB \ Perturb	$\varepsilon = 0.5$	$\varepsilon = 0.1$	$\varepsilon = 0.05$	$\varepsilon = 0.01$
Adult-UCI	68 (92%)	72 (97%)	72 (97%)	72 (97%)
Adult-Iyengar	68 (92%)	72 (97%)	72 (97%)	72 (97%)
IPUMS97-cat	97 (96%)	101 (100%)	101 (100%)	101 (100%)
IPUMS97-num	98 (97%)	101 (100%)	101 (100%)	101 (100%)

In addition to counting the nearer estimates, we also looked at the quality of the "nearness". We considered an estimate $r_{x^\star_{age}}$ to be δ-nearer w.r.t. x_{age} if

$$|r_{x^\star_{age}} - x_{age}| \leq (1-\delta)\,|x^\star_{age} - x_{age}| \ .$$

We computed the average δ of the nearer estimates. We present the results in Table 4. The higher the δ, the better the estimate. For example, a 75.11%-nearer prediction of an age is one that is within radius of 24.89% of the added noise to that age. The naive adversary is able to come to around 80% nearer, without any observable trend that depends on ε. For informed adversaries, however, it is clear that higher privacy (smaller ε) resulted in nearer estimates on average.

Table 5 shows cumulative amount of noise that was added for each database and each leakage parameter of the ε-differential perturbation, together with the amount of noise that the data contained after perturbed ages $x^\star_{age}$'s were estimated with predicted and averaged values $r_{x^\star_{age}}$'s. The naive adversary made the estimates from each sanitized data and so for each sanitization parameter ε it has different estimates and thus different amount of remaining noise. Informed and informed+expert adversaries performed the mining over one external data, and so the estimates are independent of the parameter ε.

Table 4 Average δ factor of the "nearer" estimates obtained by different adversaries

(a) Naive adversary

DB \ Perturb.	$\varepsilon = 0.5$	$\varepsilon = 0.1$	$\varepsilon = 0.05$	$\varepsilon = 0.01$
Adult-UCI	75.11%	81.10%	83.70%	84.36%
Adult-Iyengar	77.05%	76.23%	78.09%	78.70%
IPUMS97-cat	79.94%	80.78%	78.66%	81.60%
IPUMS97-num	74.45%	80.64%	81.81%	79.67%

(b) Informed adversary

DB \ Perturb.	$\varepsilon = 0.5$	$\varepsilon = 0.1$	$\varepsilon = 0.05$	$\varepsilon = 0.01$
Adult-UCI	79.09%	93.13%	96.56%	99.31%
Adult-Iyengar	79.80%	93.60%	96.80%	99.36%
IPUMS97-cat	84.49%	95.47%	97.74%	98.93%
IPUMS97-num	84.63%	96.06%	98.03%	98.95%

(c) Informed+Expert adversary

DB \ Perturb.	$\varepsilon = 0.5$	$\varepsilon = 0.1$	$\varepsilon = 0.05$	$\varepsilon = 0.01$
Adult-UCI	82.02%	94.47%	97.24%	99.45%
Adult-Iyengar	81.87%	94.65%	97.33%	99.47%
IPUMS97-cat	89.67%	96.43%	98.21%	99.64%
IPUMS97-num	92.06%	97.37%	98.69%	99.74%

Again, the smaller the leakage ε ($\varepsilon : 0.5 \rightarrow 0.1 \rightarrow 0.05 \rightarrow 0.01$), the increasingly larger percentage of noise reduction. This trend is true for informed and informed+expert adversaries, and true with a few exceptions for the naive adversary. A smaller leakage ε means higher privacy in the terms of the differential privacy notion, that is, more noise is being added to mask the ages. But the more noise is being added during perturbation, the higher fraction of the added noise is being removed by our attacks. So a major observation we make is: The higher the aimed privacy, the more successful are our attacks, which means that in the presented experimental scenarios, adding more noise does not increase privacy.

Finally, we present an example chart showing perturbed and estimated ages. Similar charts for other databases and other parameters of ε can be produced. Figure 1 shows the scenario in the case of the *IPUMS97-cat* database sanitized with perturbation satisfying ($\varepsilon = 0.5$)-differential privacy and attacked by the the most powerful informed+expert adversary. In Fig. 1(a), the x-axis consist of the original ages x_{age}'s and the y-axis shows the perturbed ages $x^{\star}_{age}$'s. The estimated ages $r_{x^{\star}_{age}}$'s from the attack are depicted in Fig. 1(b) on the y-axis. We observe that the estimates for ages between 0–60 are of good quality, especially compared to the noise seen above in Fig. 1(a). However, original ages higher that 60 could be seen as incorrectly estimated as "flat" 60–65. Overall, it is remarkable how our attack transforms the perturbed ages of Fig. 1(a) to ages in Fig. 1(b).

Table 5 Amount of noise in perturbed ages (*added noise*) and predicted ages (*remaining noise*) after perturbation and then after attacks by different adversaries

(**a**) Amount of noise added to ages during perturbation

Perturbation	**Amount of**	*Adult-UCI*	*Adult-Iyengar*	*IPUMS97-cat*	*IPUMS97-num*
$\varepsilon = 0.5$	added noise	12,316	12,316	23,970	23,970
$\varepsilon = 0.1$	added noise	61,580	61,580	119,848	119,848
$\varepsilon = 0.05$	added noise	123,160	123,160	239,695	239,695
$\varepsilon = 0.01$	added noise	615,801	615,801	1,198,477	1,198,477

(**b**) Naive adversary

Perturbation	**Amount of**	*Adult-UCI*	*Adult-Iyengar*	*IPUMS97-cat*	*IPUMS97-num*
$\varepsilon = 0.5$	remaining noise	1,767	2,137	2,976	3,556
	reduction	85.66%	82.64%	87.58%	85.16%
$\varepsilon = 0.1$	remaining noise	3,985	6,605	13,780	9,835
	reduction	93.53%	89.27%	88.50%	91.79%
$\varepsilon = 0.05$	remaining noise	7,599	12,257	29,074	18,872
	reduction	93.83%	90.05%	87.87%	92.13%
$\varepsilon = 0.01$	remaining noise	37,193	57,409	127,737	95,910
	reduction	93.96%	90.68%	89.34%	92.00%

(**c**) Informed adversary

Perturbation	**Amount of**	*Adult-UCI*	*Adult-Iyengar*	*IPUMS97-cat*	*IPUMS97-num*
	remaining noise	1,267	1,301	1,737	1,679
$\varepsilon = 0.5$	reduction	89.72%	89.44%	92.75%	93.00%
$\varepsilon = 0.1$	reduction	97.94%	97.89%	98.55%	98.60%
$\varepsilon = 0.05$	reduction	98.97%	98.94%	99.28%	99.30%
$\varepsilon = 0.01$	reduction	99.79%	99.79%	99.86%	99.86%

(**d**) Informed+Expert adversary

Perturbation	**Amount of**	*Adult-UCI*	*Adult-Iyengar*	*IPUMS97-cat*	*IPUMS97-num*
	remaining noise	1,039	1,103	1,067	864
$\varepsilon = 0.5$	reduction	91.56%	91.05%	95.55%	96.39%
$\varepsilon = 0.1$	reduction	98.31%	98.21%	99.11%	99.28%
$\varepsilon = 0.05$	reduction	99.16%	99.10%	99.55%	99.64%
$\varepsilon = 0.01$	reduction	99.83%	99.82%	99.91%	99.93%

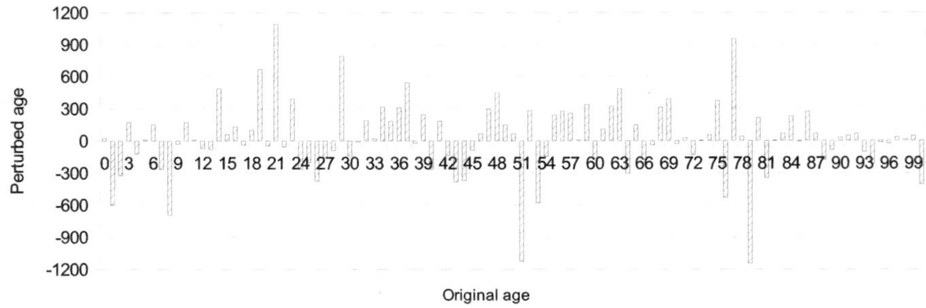

(**a**) Original and perturbed ages of *IPUMS97-cat* perturbed with $\varepsilon = 0.5$

(**b**) Original and estimated ages of *IPUMS97-cat* obtained by the informed+expert adversary

Fig. 1 Original, perturbed and estimated ages by the informed+expert adversary. *IPUMS97-cat* was sanitized with perturbation satisfying ($\varepsilon = 0.5$)-differential privacy

5 Conclusions

Data perturbation is a sanitization method that helps restrict the disclosure of sensitive information from published data. It is usually achieved by noise addition. We presented an attack which removes the majority of such added noise from the perturbed released data. The attack employs multiple data miners to make prediction about sensitive perturbed values, and it uses fusion algorithms to combine the like data and multiple predictions. Our comprehensive adversarial model captures adversaries of different strength. We explicitly modeled four types of adversaries: naive, informed, expert, and informed+expert.

The naive adversary is the weakest adversary. Its purpose it to demonstrate that our proposed attack is practical even without assuming external knowledge or expertise in data mining technologies. The consequence is that even non-expert adversaries having no deep knowledge about data mining or fusion techniques and adversaries having no additional external knowledge except the sanitized data itself can remove a remarkable amount of noise added during data perturbation and thus breach privacy of the individuals in the sanitized data. Using such naive adversaries, we were able to demonstrate negative privacy implications and vulnerabilities of data perturbation in the presented scenarios.

Stronger adversaries with unbounded external knowledge (so-called informed adversaries) or expertise in data mining field (so-called expert adversaries) are obviously able to obtain better results and make more disclosures than naive adversaries. Although unbounded adversaries are captured in our proposed adversarial model, we were able to show significant noise reduction by limiting the strengths of the adversaries in practice by selecting and fine tuning a few data mining algorithms and providing the adversary with similar individuals' data records not related to the individuals in the sanitized data.

Our overall contribution is thus twofold. We demonstrated that even simple attacks and non-expert adversaries are successful in removing noise from perturbed data, and that considering but limiting the external knowledge and expertise is enough to remove more than 99% of the noise. The consequent question is then not how to improve our attacks, but rather how to avoid them and make sensitive data publication safe from malicious parties yet useful to legal data analysts.

Our experimental results indicate that the higher the aimed privacy, the higher the percentage of noise can be removed. This suggests that adding more noise does not always increase privacy in these scenarios. The only countermeasure we have at this time is to completely mask not just the sensitive attributes of a database, but also other attributes in the data. Improper masking of relations in the records in a database results in the ability of data mining methods to make privacy-impacting predictions and estimates. On the other hand, sanitizing the data too much may render it unusable for legal data analysts. Hence, the fundamental question about balancing the amount of sanitization and the potential usefulness of the sanitized data remains an open problem.

It was already mentioned that the only countermeasure we have at this time is to completely mask not just the sensitive attributes of a database, but also other attributes in the data. As a future research topic, it would be interesting to see whether information gained from external (unsanitized) data could be used to remove a considerable part of the noise added during data perturbation not just to one attribute in the original data but to multiple or even all attributes. Another open question remains whether our proposed adversarial model and all the attacks could be used in design of new perturbation algorithms or sanitization mechanisms. For now, we foresee that our model and attacks can be used as an evaluation tool by data owners to see and evaluate the privacy risks right after they sanitize their data.

Acknowledgements. The author is with the UNESCO Chair in Data Privacy, but he is solely responsible for the views expressed in this text, which do not necessarily reflect the position of UNESCO nor commit that organization. This work was partly supported by the Spanish Government through projects TSI2007-65406-C03-01 "E-AEGIS" and CONSOLIDER INGENIO 2010 CSD2007-00004 "ARES", and by the Government of Catalonia under grant 2009 SGR 01135. Part of this work was also supported by the grant VEGA 1/0244/09, while the author was at the Slovak University of Technology, Bratislava, Slovakia.

References

1. Abowd, J.M., Vilhuber, L.: How Protective Are Synthetic Data? In: Domingo-Ferrer, J., Saygın, Y. (eds.) PSD 2008. LNCS, vol. 5262, pp. 239–246. Springer, Heidelberg (2008)
2. Adam, N.A., Wortman, J.C.: Security-control methods for statistical databases. ACM Computing Surveys 21(4), 515–556 (1989)
3. Agrawal, R., Srikant, R.: Privacy-Preserving Data Mining. In: Proceedings of the 2000 ACM SIGMOD International Conference on Management of Data, SIGMOD 2000, May 16-18, pp. 439–450. ACM Press, New York (2000)
4. Asuncion, A., Newman, D.J.: UCI Machine Learning Repository (2007), `http://www.ics.uci.edu/~mlearn/MLRepository.html`
5. Bleiholder, J., Naumann, F.: Data fusion. ACM Computing Surveys 41(1), 1–41 (2008)
6. Blum, A., Dwork, C., McSherry, F., Nissim, K.: Practical privacy: the SuLQ framework. In: Proceedings of the 24nd ACM SIGACT-SIGMOD-SIGART Symposium on Principles of Database Systems, PODS 2005, June 13-15, pp. 128–138. ACM Press, Baltimore (2005)
7. Dalenius, T.: Towards a methodology for statistical disclosure control. Statistisk Tidskrift 15, 429–444 (1977)
8. Dutta, H., Kargupta, H., Datta, S., Sivakumar, K.: Analysis of privacy preserving random perturbation techniques: further explorations. In: Proceedings of the 2003 ACM Workshop on Privacy in the Electronic Society, WPES 2003, October 30, pp. 31–38. ACM Press, Washington (2003)
9. Dwork, C.: Differential Privacy. In: Bugliesi, M., Preneel, B., Sassone, V., Wegener, I. (eds.) ICALP 2006. LNCS, vol. 4052, pp. 1–12. Springer, Heidelberg (2006)
10. Dwork, C.: Differential Privacy: A Survey of Results. In: Agrawal, M., Du, D.-Z., Duan, Z., Li, A. (eds.) TAMC 2008. LNCS, vol. 4978, pp. 1–19. Springer, Heidelberg (2008)
11. Dwork, C.: The Differential Privacy Frontier (Extended Abstract). In: Reingold, O. (ed.) TCC 2009. LNCS, vol. 5444, pp. 496–502. Springer, Heidelberg (2009)
12. Dwork, C., McSherry, F., Nissim, K., Smith, A.: Calibrating Noise to Sensitivity in Private Data Analysis. In: Halevi, S., Rabin, T. (eds.) TCC 2006. LNCS, vol. 3876, pp. 265–284. Springer, Heidelberg (2006)
13. Dwork, C., Naor, M., Reingold, O., Rothblum, G.N., Vadhan, S.P.: On the complexity of differentially private data release: efficient algorithms and hardness results. In: Proceedings of the 41st Annual ACM Symposium on Theory of Computing, STOC 2009, May 31 - June 2, pp. 381–390. ACM Press, Bethesda (2009)
14. Ganta, S.R., Acharya, R.: On Breaching Enterprise Data Privacy Through Adversarial Information Fusion. In: Proceedings of the 24th International Conference on Data Engineering Workshops, Workshop on Information Integration Methods, Architectures, and Systems, ICDE-IIMAS 2008, April 7-12, pp. 246–249. IEEE Computer Society Press, Cancun (2005)
15. Goodman, I.R., Mahler, R.P., Nguyen, H.T.: Mathematics of Data Fusion. Kluwer Academic Publishers, Norwell (1997)
16. Iyengar, V.S.: Transforming data to satisfy privacy constraints. In: Proceedings of the 8th ACM SIGKDD International Conference on Knowledge Discovery and Data Mining, KDD 2002, July 23-26, pp. 279–288. ACM Press, Edmonton (2002)
17. Kasiviswanathan, S.P., Lee, H.K., Nissim, K., Raskhodnikova, S., Smith, A.: What Can We Learn Privately? In: Proceedings of the 49th Annual IEEE Symposium on Foundations of Computer Science, FOCS 2008, October 25-28, pp. 531–540. IEEE Computer Society Press, Philadelphia (2008)

18. Machanavajjhala, A., Kifer, D., Abowd, J.M., Gehrke, J., Vilhuber, L.: Privacy: Theory meets Practice on the Map. In: Proceedings of the 24th International Conference on Data Engineering, ICDE 2008, April 7-12, pp. 277–286. IEEE Computer Society Press, Cancun (2008)
19. McSherry, F.: Preserving privacy in large-scale data analysis. A presentation at Workshop on Algorithms for Modern Massive Data Sets (MMDS 2006), Stanford, CA, USA, June 21-24 (2006), http://www.stanford.edu/group/mmds/slides/mcsherry-mmds.pdf
20. Mironov, I., Pandey, O., Reingold, O., Vadhan, S.: Computational Differential Privacy. In: Halevi, S. (ed.) CRYPTO 2009. LNCS, vol. 5677, pp. 126–142. Springer, Heidelberg (2009)
21. Muralidhar, K., Sarathy, R.: Differential Privacy for Numeric Data. In: Joint UN-ECE/Eurostat Work Session on Statistical Data Confidentiality, Bilbao, Spain (2009)
22. Muralidhar, K., Sarathy, R.: Does Differential Privacy Protect Terry Gross" Privacy? In: Domingo-Ferrer, J., Magkos, E. (eds.) PSD 2010. LNCS, vol. 6344, pp. 200–209. Springer, Heidelberg (2010)
23. Sarathy, R., Muralidhar, K.: Some Additional Insights on Applying Differential Privacy for Numeric Data. In: Domingo-Ferrer, J., Magkos, E. (eds.) PSD 2010. LNCS, vol. 6344, pp. 210–219. Springer, Heidelberg (2010)
24. Sramka, M.: A Privacy Attack That Removes the Majority of the Noise From Perturbed Data. In: Proceedings of the 2010 International Joint Conference on Neural Networks, IJCNN 2010, as part of the 2010 IEEE World Congress on Computational Intelligence, WCCI 2010, July 18-23. IEEE Computer Society Press, Barcelona (2010)
25. Sramka, M.: Data mining as a tool in privacy-preserving data publishing. Tatra Mountains Mathematical Publications 45, 151–159 (2010)
26. Sramka, M., Safavi-Naini, R., Denzinger, J.: An Attack on the Privacy of Sanitized Data That Fuses the Outputs of Multiple Data Miners. In: Proceedings of the 9th IEEE International Conference on Data Mining Workshops, International workshop on Privacy Aspects of Data Mining, ICDM-PADM 2009, December 6, pp. 130–137. IEEE Computer Society Press, Miami Beach (2009)
27. Sramka, M., Safavi-Naini, R., Denzinger, J., Askari, M.: A Practice-oriented Framework for Measuring Privacy and Utility in Data Sanitization Systems. In: Proceedings of the 12th International Conference on Extending Database Technology Workshops, the 3rd International Workshop on Privacy and Anonymity in the Information Society, EDBT-PAIS 2010, March 22-26. ACM Press, Lausanne (2010)
28. Sramka, M., Safavi-Naini, R., Denzinger, J., Askari, M., Gao, J.: Utility of Knowledge Extracted from Unsanitized Data when Applied to Sanitized Data. In: Proceedings of the 6th Annual Conference on Privacy, Security and Trust, PST 2008, October 1-3, pp. 227–231. IEEE Computer Society Press, Fredericton (2008)
29. Torra, V. (ed.): Information Fusion in Data Mining. Studies in Fuzziness and Soft Computing, vol. 123. Springer, Heidelberg (2003)
30. Torra, V., Narukawa, Y.: Modeling Decisions: Information Fusion and Aggregation Operators. In: Cognitive Technologies. Springer, Heidelberg (2007)
31. Valls, A., Torra, V., Domingo-Ferrer, J.: Semantic based aggregation for statistical disclosure control. International Journal of Intelligent Systems 18(9), 393–951 (2003)
32. Witten, I.H., Frank, E.: Data Mining: Practical machine learning tools and techniques, 2nd edn. Morgan Kaufmann, San Francisco (2005)

Feature Selection of Hand Biometrical Traits Based on Computational Intelligence Techniques

R.M. Luque, D. Elizondo, E. López-Rubio, and E.J. Palomo

Abstract. This chapter presents a novel methodology for using feature selection in hand biometric systems, based on genetic algorithms and mutual information. The aim is to provide a standard features dataset which diminishes the number of features to extract and decreases the complexity of the whole identification process. The experimental results show that it is not always necessary to apply sophisticated and complex classifiers to obtain good accuracy rates. This methodology approach manages to discover the most suitable geometric hand features, among all the extracted data, to perform the classification task. Simple classifiers like K-Nearest Neighbour (kNN) or Linear Discriminant Analysis (LDA) in combination with this strategy, getting even better results than other more complicated approaches.

1 Introduction

Nowadays, identification systems based on biometric information are a reality which is crossing from research into industry. Because of the advances in developing hardware at low cost, computer-aided personal recognition and identification technology with biometrics features belonging to iris, fingerprint, face, voice or hands, has become more and more relevant because of its stability and reliability [1].

D. Elizondo
Department of Computer Technology, Forsec, De Monfort University
The Gateway, LE1 9BH – Leicester, United Kingdom
e-mail: {elizondo}@dmu.ac.uk

R.M. Luque · E. López-Rubio · E.J. Palomo
Department of Computer Languages and Computer Science, University of Málaga
Campus Teatinos s/n, 29071 – Málaga, Spain
e-mail: {rmluque, ezeqlr, ejpalomo}@lcc.uma.es

D.A. Elizondo et al. (Eds.): Computational Intelligence for Privacy and Security, SCI 394, pp. 159–180.
springerlink.com

Although the use of biometrical systems has been extended to a large number of applications, this does not imply that biometrics is a fully solved problem or that no additional research is required. Hand geometry recognition systems are getting a considerable importance in medium security applications because they take advantage of several factors which other biometrics traits do not [2]. From the hardware viewpoint, just low resolution cameras are needed, and the computational cost of the applied algorithms is relatively scarce, so it is easily accepted by the users.

Hence, hand-based biometric systems have attracted and motivated the interest of a large number of researchers for the last years [3, 4]. Two different groups of features have usually been extracted to perform the identification process. Geometric-based hand features have been used in earlier papers [2] since they are easily computed and provide acceptable prediction rates over small-medium hand data sets (no more than 50-100 users). Other approaches are based on texture features which are extracted from robust and invariant palmprint hand region [5, 6, 7]. However, the number of features is significantly increased, depending on how many coefficients for every single texture method are selected to represent the Region of Interest (ROI).

Several papers have also exploited the combination of the two previous groups of features in order to increase the reliability and accuracy of the biometric systems [8, 9, 10]. Different schemes for feature fusion have been proposed [11]. Basically, they depend on the level at which the feature combination is performed, namely: feature extraction, matching score and decision. Since in biometric applications we have just a few individual patterns for each user and a huge amount of features obtained after the image processing performance, the curse of dimensionality problem should be taken into account to avoid low accuracy rates due to a lack of generalization in the classifiers.

For this reason, feature selection is considered one of the main difficult tasks in hand biometric systems: to find out what the most significant variables are and whether they provide enough information to conduct an efficient classification task. Therefore, although it is possible to extract a large amount of hand features, this does not ensure the best accuracy rate, since many of them could be superfluous or even useless and the classifier might not be able to generalise correctly.

There are two different alternatives in the literature to deal with this task. The simpler one extracts just the most discriminant features. This implies a prior knowledge of the dataset analysed. This is interesting when advise from an expert in biometrics [12] with a wide knowledge on the problem domain is available. In that case, it is better to construct a set of *ad hoc* features. The second one involves that, in most cases, it is not known how important and discriminant the variables are, so statistical feature selection techniques such as Principal Components Analysis (PCA), Independent Components Analysis (ICA) or Linear Discriminant Analysis (LDA) [13, 14, 15, 9] are applied to reduce the dimensionality of the data, which also cause a decrease in the complexity of the subsequent classifier.

1.1 Motivation

Most of the previous studies lack a justification to extract one set of features over another. Many of them represent the biometrical hand traits by using a specific set of features which is chosen without applying any selection criterion. They then attempted to prove if these variables, or any linear or non-linear combination of them, are useful as inputs to the prediction algorithm. Consequently, there have been more efforts towards building efficient and accurate classifiers than in the creation of standard and well-defined subsets of discriminant features.

Common feature selection methods such as PCA or ICA have turned out quite effective and useful in reducing the number of inputs and improving the prediction performance of the predictors. However, they generate a new dataset with less features, but which is different from the initial one. Thus, they represent the data in a lower dimensional space computed through a linear or non-linear transformation, which does not provide a good understanding of which of the initial features are the most relevant. Furthermore, these methods do not ensure that the resulting subset of features is minimal nor if the extracted subset of features will be suitable to obtain better classification rates.

Genetic algorithms (GA) are considered as suitable evolutionary strategies for feature selection in the literature. They are well adapted for problems with a large number of features [16, 17, 18], and are applied to different areas, from object detection [19] to gene detection in microarray data [20]. With this kind of feature selection approach it is expected to achieve a three-fold objective: improving the accuracy rate of the classifiers; providing faster and more cost-effective predictors because of a significant reduction of the number of features; and providing a better understanding of the underlying process that generated the data.

In our approach, the GA is employed in combination with two simple classifiers, the k nearest neighbour classification rule (kNN)[21] and the Linear Discriminant Analysis (LDA)[22], in order to analyse the benefits of our evolutionary approach with the least possible influence of the employed classifier.

Since several features might be correlated, it is possible that each iteration of the genetic algorithm generates a different suitable feature subset. This would mean that there are different sets of feature subsets whose accuracy rate is optimal. Therefore, it would be relevant to find out the relationship between the features of two optimal feature subsets, in order to determine if they provide the same type of information. We combine computational intelligence techniques with statistical approaches such as mutual information [23], with the aim of providing feedback about whether or not the subsets generated are equivalent in terms of the correlation among the variables of each subset.

The structure of the chapter is as follows: Sections 2 and 3 describe steps needed to achieve robust and reliable geometric hand features; Section 4 sets out the feature selection methodology to provide a powerful and discriminant features subset; Section 5 describes the classification methods that are applied to check the viability of the previous subset; finally, in Section 6 some experimental results are showed; conclusions are outlined in Section 7.

2 Image Preprocessing

Image preprocessing techniques are very dependent on the quality and characteristics of the available datasets, and a great number of factors can influence the final result. In fact, different possibilities in the colour space (RGB or grey-scale), the contrast between the foreground and background, the position of the hand in the images, the clarity of the boundaries definition, the appearance of positioning aids in the system [2], etc. should be taken into account to develop a preprocessing phase which can ensure a stable and robust hand segmentation, in order to extract the required features as accurately as possible. Therefore, the main aim of this stage is to provide a reasonable starting point to the feature extraction process.

It should be noted that the more constraints the design of the physical system has, the simpler the preprocessing phase will be. Thus, if it is possible to avoid sudden illumination changes and achieve homogeneous lighting during the acquisition process, the image binarisation task will be less complex and arduous. If the acquisition system allows to place the hand with certain freedom, this task will not be too tiresome for the end-user. Although it is necessary that some basic rules are fulfilled, the benefits in terms of less complexity of the positioning algorithm will raise. Consequently, these physical requirements considerably increase the reliability and efficiency of the process in its early stages. They will also improve the accuracy of the feature extraction and the robustness and viability of the recognition system.

Since we have used the GPDS Hand database[1] acquired from the University of Las Palmas de Gran Canaria [24], these constraints are already defined. Thus, the images should be analysed in order to conduct a suitable preprocessing procedure. This database is made of 50 individuals with 10 different captures of their top right hand. The 500 images have been taken by using a desk-scanner without pegs or annoying templates which can hinder placing the hand correctly. The male hand images constitute about 68% of all data, whereas the rest are associated to women. The age of the users varies from 23 to 30. The aim is to utilise hand images from similar ages in order to have the worst possible situation in which the visual shape variation is not so significant. The images have a resolution of 150 dpi and 256 grey levels.

The subsequent subsections present in more detail, the involving steps of the proposal image preprocessing approach. Figure 1 shows graphically the whole process. Extracting a region of interest (ROI) from the centre of the hand image to get texture-based features is not needed, since we are interested in geometric features detection only.

2.1 *Binarisation*

After analysing the images from GPDS dataset, it is noticed that the background image is largely black and there is a sufficiently high contrast with regard to the foreground. The lighting is homogeneous in the entire image and there is no abrupt

[1] The database can be downloaded in http://www.gpds.ulpgc.es/download/index.htm

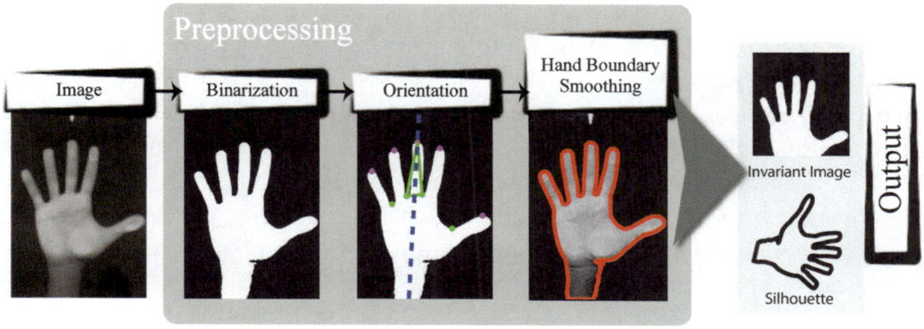

Fig. 1 Stages of the preprocessing module. The outputs are used as the starting point to extract the required features.

illumination changes from one image to another. Nevertheless, a median-based filter is applied to smooth the peaks and reduce the unpleasant pixel variations.

Provided that the image background is stable, an image thresholding operation is used to obtain a binary hand-shape image. The threshold value is automatically computed by using the Otsu's method [25], and is weighted with a fixed rate (between 0 and 1) obtained empirically according to the database. Subsequently, morphological operators are applied to clean the image, by filling holes and removing spurious objects.

2.2 *Orientation*

Since the acquisition system is free of guidance pegs, the hand can be positioned anywhere within the area of the scanner window and it is necessary to locate the region which is associated to the hand. Alignment rules are used to guide the users to put the hand in a more suitable way could be implemented. However, a rotational alignment mechanism should be incorporated to achieve better robustness.

Furthermore, on several occasions the binarised image obtained after the acquisition process consists of the union of the hand and part of the arm. In [26], an iterative morphological closing algorithm using a circular structural element is presented to find the correct segmentation. Its main drawback is that it is time consuming, although they implement a strategy based on multi-resolution to reduce this complexity.

The proposed alternative [27, 13] discards the part of the binary image which is not related to the hand, by using a hand contour algorithm to detect the key points of the hand. Therefore, a silhouette of the hand is obtained by applying a chain code technique over the binary image. Chain codes are used to represent a boundary by a connected sequence of numbers based on the direction of the following pixel in the contour. Typically, this representation is based on 4- or 8- connectivity of the pixels, in which each direction is coded by using a numbering scheme. The chain code of a boundary is very dependent on the starting point, so the meaning of this point has to always be the same in this approach. In this particular case, the inferior left pixel edge of the binary hand was chosen. Preliminary key points have been selected by using the same process that is applied in [12].

The middle finger line is computed and any rotation is considered with reference to it. The use of elliptical approaches [5] for aligning the hand fails because of the large part of the wrist imaged. Finally, an invariant binary hand region is obtained from the valley between the index and thumb fingers, taking a vertical line and getting the bottom edge hand pixel, and taking a horizontal line from this point to get the superior binary part.

2.3 Hand Boundary Smoothing

The binarisation process is not always so robust and accurate as expected. This implies the hand edges could present some noise and imperfections, which alter the geometric measurements to compute. If the contour is considered as a parametric curve $(x(t), y(t))$, it can be smoothed by applying a 1D kernel regression technique. In this case, x is the x-coordinate of the contour pixel, y is the y-coordinate of the contour pixel and t is position of each pixel in the curve, e.g. $t = 1$ for the first pixel of the contour, and so on.

Let t be an arbitrary position in the contour; it could be the case that t is not an integer, i.e. subpixel accuracy is allowed. The local kernel estimators in the vicinity of t are given by [28]:

$$x(t_i) = x(t) + \frac{\partial x(t)}{\partial t}(t_i - t) + \frac{1}{2}\frac{\partial^2 x(t)}{\partial t^2}(t_i - t)^2 + \ldots \tag{1}$$

$$y(t_i) = y(t) + \frac{\partial y(t)}{\partial t}(t_i - t) + \frac{1}{2}\frac{\partial^2 y(t)}{\partial t^2}(t_i - t)^2 + \ldots \tag{2}$$

The parameters to be determined are:

$$\mathbf{b}_x = \left[x(t), \frac{\partial x(t)}{\partial t}, \frac{\partial^2 x(t)}{\partial t^2}\right] \tag{3}$$

$$\mathbf{b}_y = \left[y(t), \frac{\partial y(t)}{\partial t}, \frac{\partial^2 y(t)}{\partial t^2}\right] \tag{4}$$

The main interest is in the values of $x(t)$ and $y(t)$. The parameter vectors $\mathbf{b}_x$ and $\mathbf{b}_y$ are computed by solving the following two optimisation problems:

$$\hat{\mathbf{b}}_x = \arg\min_{\mathbf{b}_x} F(\mathbf{b}_x) \tag{5}$$

$$F(\mathbf{b}_x) =$$

$$\sum_i K_i(t_i - t)\left(\hat{x}_i - x(t) - \frac{\partial x(t)}{\partial t}(t_i - t) - \frac{1}{2}\frac{\partial^2 x(t)}{\partial t^2}(t_i - t)^2 - \ldots\right)^2 \tag{6}$$

$$\hat{\mathbf{b}}_y = \arg\min_{\mathbf{b}_y} F(\mathbf{b}_y) \tag{7}$$

$$F(\mathbf{b}_y) = \sum_i K_i(t_i - t)\left(\hat{y}_i - y(t) - \frac{\partial y(t)}{\partial t}(t_i - t) - \frac{1}{2}\frac{\partial^2 y(t)}{\partial t^2}(t_i - t)^2 - \ldots\right)^2 \tag{8}$$

where the summation on i runs over a smoothing interval centered on t, $\hat{x}_i$ and $\hat{y}_i$ are the original (unsmoothed) values at position i, and K_i is the smoothing kernel function for position i,

$$K_i(t_i - t) = \frac{1}{\sqrt{2\pi\rho^2}} \exp\left(-\frac{(t_i - t)^2}{2\rho^2}\right) \tag{9}$$

with ρ being a global smoothing parameter. The result of this procedure is the smoothed version of the parametric curve, $(x(t), y(t))$.

3 Biometrical Feature Extraction

In this section the set of extracted features is described. Only geometric hand features excluding texture features or a combination of both were considered. The main reasons for this choice are:

- Geometric features are simpler to compute and have proved their usefulness in biometric recognition problems [12, 29].
- Unlike texture features which are a set of coefficients from a ROI, their meaning is more visual and understandable since they correspond to physical measurements.
- Classifying hand sets which are made of between 50-100 users is an affordable task which does not require high complexity or resources.
- It is possible to get a large number of geometric features by just applying the corresponding techniques to extract them.

By using the previous data after the preprocessing phase, the most common geometric features were selected, in order to generate a standard set of discriminant features among all existing possibilities. Next, the descriptors which extract the features are described. All of the features are applied both to the hand and to each finger individually (Figure 2).

The *area* descriptor (A) is computed by summing all the active pixels of region (hand or finger), whereas the *perimeter* (p) is based on the length of the chain code. It is calculated by summing the horizontal and vertical pixels and the diagonal ones weighted by $\sqrt{2}$, which is the hypotenuse of a triangle whose cathetus are both one horizontal pixel and one vertical one. As well as the *width* and *height* descriptors, the *aspect proportion* is included. This is computed by dividing the width and the

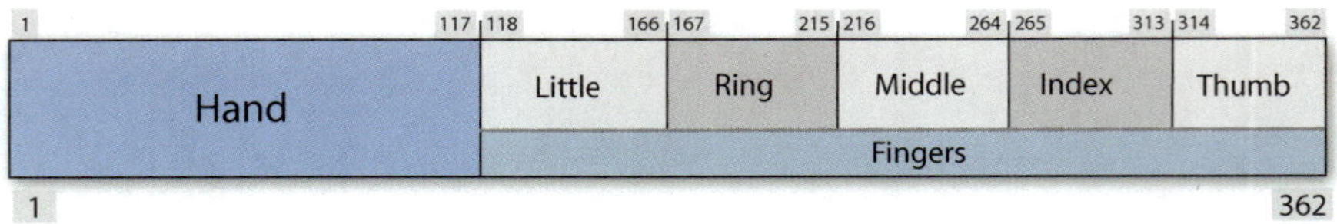

Fig. 2 Vector features distribution. A third of the vector corresponds to hand features whereas the rest is associated to fingers features. The numbers represent the number features which delimit the start and end of each block.

height. In the case of the fingers, three equidistant width measures are taken along the main direction.

The *compactness* ratio of a region is defined as $c = 4\pi(A/p^2)$, and the *solidity* can be easily computed after getting the convex area [30] as $s = A/Convex_A$. Image moments are important statistical properties of an image and their usefulness have been proved to describe objects after segmentation [31]. A set of seven invariant moments, called *Hu moments*, which are insensitive to translations, scale changes, rotations and transformations mirror are calculated and given by the following equations:

$$
\begin{aligned}
\phi_1 &= \eta_{20} + \eta_{02} \\
\phi_2 &= (\eta_{20} - \eta_{02})^2 + 4\eta_{11}^2 \\
\phi_3 &= (\eta_{30} - 3\eta_{12})^2 + (3\eta_{21} - \eta_{03})^2 \\
\phi_4 &= (\eta_{30} + \eta_{12})^2 + (\eta_{21} + \eta_{03})^2 \\
\phi_5 &= (\eta_{30} - 3\eta_{12})(\eta_{30} + \eta_{12}) + [(\eta_{30} + \eta_{12})^2 - 3(\eta_{21} + \eta_{03})^2] + \\
&\quad (3\eta_{21} - \eta_{03})(\eta_{21} + \eta_{03})\,[3(\eta_{30} + \eta_{12})^2 - (\eta_{21} + \eta_{03})^2] \\
\phi_6 &= (\eta_{20} - 3\eta_{02})\,[(\eta_{30} + \eta_{12})^2 - (\eta_{21} + \eta_{03})^2] + \\
&\quad 4\eta_{11}(\eta_{30} + \eta_{12})(\eta_{21} + \eta_{03}) \\
\phi_7 &= (3\eta_{21} - 3\eta_{03})(\eta_{30} + \eta_{12})\,[(\eta_{30} + \eta_{12})^2 - 3(\eta_{21} + \eta_{03})^2] + \\
&\quad (3\eta_{12} - \eta_{03})(\eta_{21} + \eta_{03})\,[3(\eta_{30} + \eta_{12})^2 - (\eta_{21} + \eta_{03})^2]
\end{aligned}
$$

It should be noted that these invariant measures are not applied to the fingers as there are very little differences between classes.

A closed silhouette defined as $(x(t), y(t))$ can be represented in terms of Fourier descriptors, where t is the length of the curve from the starting point. After selecting a number of equidistant points, N, of the curve

$$(x(t), y(t)), \quad n = 0, 1, 2, \ldots, N-1$$

it is possible to obtain the complex vector

$$z = x + iy$$

where $x = (x(0), x(1), \ldots, x(N-1))$ and $y = (y(0), y(1), \ldots, y(N-1))$.

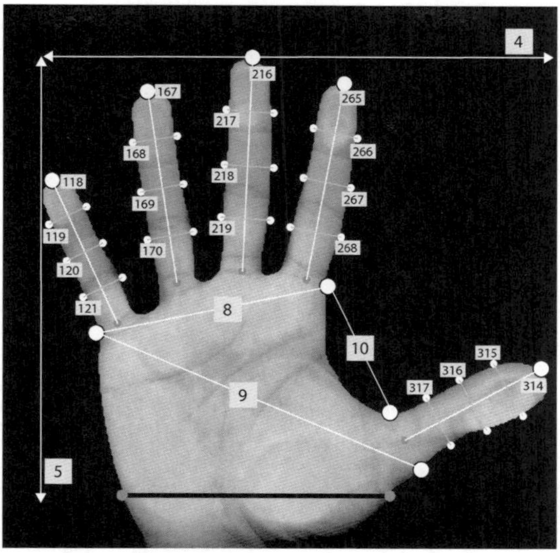

Fig. 3 Representation of the extracted features. Only the visual ones are showed. The meaning of each feature is commented in Table 1.

The Fourier discrete transform of this vector by using just the first P Fourier descriptors is described as follows

$$\hat{Z}(u) = \frac{1}{N}\sum_{u=0}^{P-1} z(n)exp(-\frac{2\pi nui}{N})$$

where $u = 0,1,2,\ldots,N-1$ and $\hat{Z}(u) = 0, u = P+1,\ldots,N-1$ because $\hat{Z}$ is an approximation. Both the magnitude $|Z(k)|, k = 0,1,\ldots,P-1$ and the phase $arg(Z(k)), k = 0,1,\ldots,P-1$ of the Fourier descriptors are used as inputs in the feature vector. The hand silhouette is presented by using more Fourier descriptors ($P = 50$) than the fingers silhouette ($P = 20$).

Finally, three robust distances between interior points of the hand contour are computed [12]. These features together with the rest of the visible ones are represented in Figure 3. A total of 362 features were extracted from each hand. Table 1 describes all the extracted features and their position in the feature vector.

4 Feature Selection Methodology

This section presents the feature selection methodology to find out the subset of features which provide high accuracy rates in the classification task. Many classical approaches such as PCA or ICA cast the data in a lower dimensional space through a linear or non-linear transformation. However, they generate a new dataset with less features, but which is different from the initial one, not providing a good understanding of which of the initial features are the most relevant.

Table 1 Meaning of each feature detected. Each number is associated to the descriptor in the first column. The column of the feature number indicates the place into which the descriptor is applied. Characteristic numbers correspond to the measurements which are shown in Figure 3.

	Characteristics					
	Hand	*Fingers*				
Descriptor		*Little*	*Ring*	*Middle*	*Index*	*Thumb*
Area	1	122	171	220	269	318
Perimeter	2	123	172	221	270	319
Compactness	3	124	173	222	271	320
Width	4	119-121	168-170	217-219	266-268	315-317
Length	5	118	167	216	265	314
Aspect Proportion	6	126	175	224	273	322
Solidity	7	125	174	223	272	321
Distance Little-Index	8	-	-	-	-	-
Distance Little-Thumb	9	-	-	-	-	-
Distance Index-Thumb	10	-	-	-	-	-
Fourier Magnitude	11-60	147-166	196-215	245-264	294-313	343-362
Fourier Phase	61-110	127-146	176-195	225-244	274-293	323-342
Hu Moments	111-117	-	-	-	-	-

Genetic algorithms (GA) are applied to select the minimum subset of initial features which ensure high results. The framework of the proposed methodology is showed in Figure 4. Although this can be a time-consuming technique (depending on the classifier), the GA is applied only to the training stage of the system, reducing resources and time complexity. In addition to that, the selected subset is minimised and it gets high prediction rates.

Therefore, the main objective is to select the most relevant variables to define a standard set of features for hand biometric systems. It should be noted that this subset is largely dependent in the dataset utilised. The environment conditions and the image quality in other datasets could change and produce another subset. Nevertheless, this feature extraction methodology can be extrapolated to different datasets, with no limitation in the kind of descriptors to apply.

4.1 Evolutionary Strategy

Genetic algorithms [16, 17, 18] are a class of optimisation procedures inspired by the biological mechanisms of reproduction. In this kind of optimisation problems, a

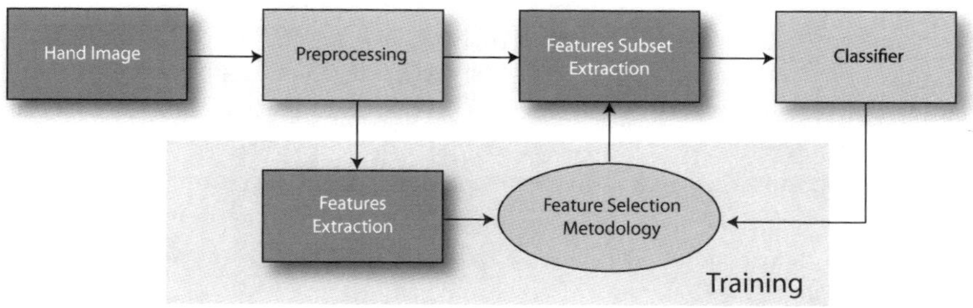

Fig. 4 Framework of our approach. The training phase selects the features subset with best accuracy to use it as the standard feature subset to extract.

fitness function $f(x)$ should be maximised or minimised over a given space X of arbitrary dimension. In this case, the fitness function combines the aim of minimising the number of features and the error classification rate. If the number of variables is not too large, an exhaustive search could be suitable, but in optimisation problems with a considerable amount of features, this is unfeasible.

Three basic genetic operators guide this search from an initial population: selection, crossover, and mutation. The genetic search process is iterative: evaluating, selecting, and recombining strings in the population during each (generation) until reaching some termination condition. Two stopping criteria are considered. The first one is the maximum number of generations, whereas the second one is the number of generations in which no improvement is obtained. In this approach, the first number is assigned to 300 and the second to 50. In most cases, the GA converges in less than 300 generations.

4.1.1 Encoding and Initial Population

A simple encoding scheme to represent as much as possible the search space is employed, in which the chromosome is a bit string whose length is determined by the number of features extracted. Each variable is associated with one bit in the string. If the ith bit is active (value 1), then the ith feature is selected in the chromosome. Otherwise, a value of 0 indicates that the corresponding feature is ignored. Each chromosome represents a different feature subset to evaluate.

Generally, the initial population is generated randomly, where each bit in the chromosome is set to 0 or 1 with a fifty percent of probability. However, this process produces a population in which all the chromosomes contain approximately the same number of 1's and 0's on average. Therefore, to analyse subsets with different number of features, both the number of active features and the choice of what features are active, are generated randomly. In all of our experiments, we used a population size of 100 individuals.

4.1.2 Selection

The selection function chooses parents for the next generation by favouring fitter candidates over weaker ones. A strategy based on roulette wheel and uniform sampling was applied [32], which involves mapping the individuals in contiguous segments of a line, such that each individual segment is equal in size to its fitness value.

Furthermore, it is rather advisable to retain the chromosomes with the best fitness values for the next generation. These individuals are called elite children. In these experiments, an elite count value of 10 is selected.

4.1.3 Crossover and Mutation

Scattered crossover, in which each bit of the offspring is selected randomly from the corresponding bits of the parents, is the choice for combining parents of the previous generation. The crossover rate is established to 0.8 in this approach.

In addition to that, a traditional mutation operator which flips a specific bit with a probability rate of 0.2 is considered. A modification which involves mutating a random number of bits between 1 and the number of active features of the individual is introduced. Since it was empirically checked that the best subsets include few features, this change avoids the increment on the number of active features in the last generations of the GA.

4.1.4 Fitness Function

The fitness function assesses each chromosome in the population so that it may be ranked against all the other chromosomes. The main goal of feature subset selection is to use less features to achieve the same or better performance. Therefore, the fitness function should contain two terms: the accuracy and the number of features selected. To evaluate how accurate the proposed subset is in terms of prediction, the dataset is split into training and testing sets. Although any supervised classifier can be used to assess the suitability of the subset, two of the simplest ones have been chosen and described in Section 5.

Thus, the classifier is trained using the first set and validated using the second one. The percentage of data in each set was weighted by 0.3 for training and 0.7 for testing. Since ten hand images are obtained for each user, just three of them are used to train the classifier whereas the remaining seven images are taken into account in the assessment.

The function to be minimised is represented as follows:

$$fitness(x) = (1 - accuracy(x)) + \lambda \frac{no.\ active\ features}{no.\ total\ features} \tag{10}$$

where $fitness(x)$ is the fitness value of the feature subset represented by x; $accuracy(x)$ is the level of generalisation obtained by the classifier using the validation set; finally, the $no.\ active\ features$ and $no.\ total\ features$ are the number of active bits in the individual and its total length, respectively. The λ value is in the interval $(0,1)$ and is assigned to $\lambda = 0.8$.

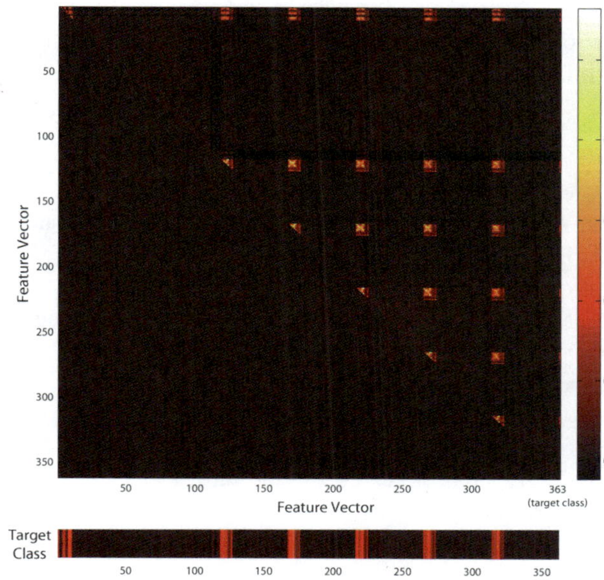

Fig. 5 Pairwise feature correlation using mutual information. The lighter the colour, the more correlated the features. It is easily remarkable the number of features clusters correlated. The target class labels are included as the feature 363 (more details in the bottom bar). The most discriminant features are correlated with this variable.

Therefore, if two subsets achieve the same performance, while containing different number of features, the subset with fewer features is preferred. Between accuracy and feature subset size, accuracy is our major concern.

After running the GA several times, it is common to find some feature subsets which provide similar accuracy values. These subsets are considerably different, so it is necessary to find out if the variables of the two subsets are correlated.

4.2 Mutual Information

Statistical techniques such as mutual information [33, 34] give us an idea of the correlation between a pair of features. As a result, several clusters of correlated features could arise. In this case, all the obtained subsets would be permutations of the features of each cluster discovered by measuring the mutual information. It is important to note that the selected features by the genetic algorithm largely depend on the classification method used as part of its fitness function.

The mutual information (in bits) between two continuous random variables x and y is given by

$$I(x,y) = \int \int p(x,y) \log_2 \left(\frac{p(x,y)}{p(x)\,p(y)} \right) dx dy \tag{11}$$

where $p(x,y)$ is the joint probability density function of x and y, and $p(x)$ and $p(y)$ are the marginal probability density functions of x and y respectively. Please note that the mutual information is symmetric:

$$I(x,y) = I(y,x) \tag{12}$$

Moreover, it is non-negative, with a zero value indicating that the variables are independent. The more correlated two variables are, the greater their mutual information.

In a classification problem such as the present one, the main interest in the mutual information between pairs of features (to avoid redundancies in the set of selected features), and in the mutual information between each feature and the class variable (to maximise the relevancy of the selected features). The relationships between pairs of variables are best visualised by plotting a matrix with all the pairwise mutual information [35], as seen in Figure 5. After analysing the matrix, six compact clusters of features can be observed. The features in the same group are strongly related to each other and also related to the target class variable, which is represented in the bottom part of the figure.

Therefore, it might be enough to select one feature from each one of the six clusters. However, the behaviour of the classifier with these features does not ensure the highest accuracy rate. Both redundant and useless variables can improve the result in some occasions [23] getting better rates in the prediction. The combination of mutual information and the GAs is analysed in Section 6.

5 Classification Methods

Two simple supervised classification methods are selected as part of the fitness function of the genetic algorithm. These two methods are the k-nearest neighbour algorithm (kNN) and the linear discriminant analysis (LDA), which are simple linear classifiers since the approach discussed is more oriented to the feature selection strategy that the classification task itself.

5.1 K-Nearest Neighbour Algorithm

The k-nearest neighbours algorithm (kNN)[21] classifies a test pattern by assigning the class of the one of the k closest training patterns for which the class is known, given a distance measure. The most standard measure is the Euclidean distance, although other metrics (e.g., Mahalanobis distance) are sometimes used. If $k = 1$, then the pattern is simply assigned to the class of its nearest neighbour. Otherwise, there is a majority vote strategy among the k closest patterns to decide the final class.

The kNN classification rule has been selected for use in combination with the GA mainly because of its simplicity. Furthermore, it is a really fast algorithm specially compared to other supervised classifiers like Support Vector Machine (SVM) or Multi-layer Perceptron (MLP), and it is very powerful provided that a good distance measure and a suitable k value are found. Finally, this classifier is well explored in the literature, and has been demonstrated to have good classification performance on a wide range of real-world data sets.

5.2 Linear Discriminant Analysis

Linear discriminant analysis (LDA)[22] is a method used in pattern recognition and machine learning to find a linear combination of features which separate two or more classes of patterns. It is used both as linear classifier as for dimension reduction of a set of input features.

In its classification form, its aim is to construct a set of linear functions of the predictors, known as discriminant functions, maximising the ratio of among-groups to within-groups variance. If data is described by a matrix $X = (x_1, x_2, \dots, x_n)$ where n is the number of patterns and each x_i vector is made of q features, μ_k is the mean of the class $\mathscr{C}_k$ and μ is the mean of all the patterns, then:

$$S_B = \sum_k (\mu_k - \mu)(\mu_k - \mu)^t \qquad S_W = \sum_k \sum_{x_i \in \mathscr{C}_k} (x_i - \mu_k)(x_i - \mu_k)^t \tag{13}$$

where S_B and S_W represent the between-class and within-class covariances respectively. The purpose of the LDA method is to get a vector w which tries to increase the between-class separation by maximising $w^t S_b w$, and decrease the within-class separation by minimising $w^t S_W w$. This double objective is satisfied by the vector w_{opt} which maximises the following criterion:

$$J(w) = \frac{w^t S_b w}{w^t S_W w} \tag{14}$$

If the inverse of S_W exists, then a solution w_{opt} can be found.

6 Experimental Results

In this section, the results of our approach are showed and analysed. The GPDS-hand database with 50 individuals and 10 samples for each one is used in all the experiments. In total, a dataset of 500 images is available, in which 150 images have been used for training whereas 350 have been used for validation (30% and 70% respectively). All the experiments reported on this chapter have been carried out with Matlab on a 32-bit PC with a quad core 2.40GHz CPU and 3GB RAM.

Two experiments have been performed using the GA methodology. The main difference is the classifier applied in the fitness function of the GA, the linear discriminant analysis (LDA) in the first case and the k-nearest neighbours (kNN) in the second one. Since it is likely to find a large number of subset solutions with the same accuracy, the GA is performed several times with each classifier. Thus, 120 different executions of the GA have been carried out, in which 75 of them corresponds to the GA-LDA strategy and the remaining 45 uses the GA-kNN one. Table 2 shows the

Table 2 Accuracy results from different feature subsets. The first column shows the variables in each subset corresponds to each execution of the GA. The fitness function is defined in terms of the level of generalisation obtained by two classifiers (third column). The fourth and fifth columns are the results obtained with kNN and LDA classifiers. They correspond to the accuracy rates (mean±std) using the features subset (first column) as the inputs of the classifiers.

			Results	
Variables	*No.Features*	*Classifier*	*kNN*	*LDA*
5, 122, 168, 265, 266	5	kNN	**99.714±0.17**	99.703±0.34
169, 216, 219, 273, 314, 316	6	kNN	99.931±0.15	**99.949±0.16**
5, 10, 126, 217, 219, 224	6	kNN	99.743±0.29	**99.800±0.28**
8, 118, 218, 220, 265, 314	6	kNN	99.709±0.22	**99.874±0.16**
8, 9, 121, 167, 217, 265, 273, 317	8	kNN	**99.857±0.18**	99.794±0.22
119, 126, 216, 218, 265, 266	6	kNN	**99.966±0.11**	99.931±0.14
118, 119, 168, 216, 224, 269	6	kNN	**99.869±0.18**	99.851±0.21
9, 126, 216, 217, 224, 269, 318	7	kNN	**99.771±0.23**	99.646±0.23
5, 6, 168, 173, 267, 269, 317	7	LDA	94.006±1.08	**99.897±0.19**
5, 8, 119, 218, 273, 318	6	LDA	99.777±0.34	**99.926±0.15**
118, 119, 216, 224, 265	5	LDA	99.606±0.26	**99.891±0.17**
10, 126, 216, 266, 314	5	LDA	99.126±0.50	**99.869±0.17**
8, 122, 126, 216, 266	5	LDA	99.457±0.37	**99.840±0.18**
5, 169, 218, 219, 269, 314	6	LDA	99.629±0.33	**99.817±0.25**
118, 119, 169, 216, 224, 269	6	LDA	99.806±0.25	**99.891±0.17**
167, 171, 218, 219, 269, 314	6	LDA	98.766±0.51	**99.663±0.25**
5, 175, 265, 269, 314, 317	6	LDA	99.200±0.46	**99.811±0.21**

results of the GA performance. It should be noted the important reduction of the number of features, from an initial set of 362 variables to feature subsets with no more than eight features in all the executions. This decrease does not involve a lack of quality in the classification, since we get accuracy rates which are well above 99.5%.

The relation between the final feature subset chosen and the classifier used in the fitness function (LDA or kNN) is also taken into account. Thus, it is easily observable that slightly better results are obtained if the classifier is the same both in the feature selection and classification stages. Therefore, the choice of the features in the GA strategy depends on the classifier. Nevertheless, the differences between the two classification techniques are not significant, which implies the final chosen subsets are completely valid for the two studied methods.

The relationship between different optimal feature subsets is also identified. In fact, Figure 6 shows correlation amongst the variables of all the subsets by using the mutual information method. Each horizontal line corresponds to the subset solution of one GA execution. Furthermore, six significant clusters of features are observed (shaded colour). They are associated to features of the hand and the five fingers respectively (from left to right). Two variables of different subsets are correlated if

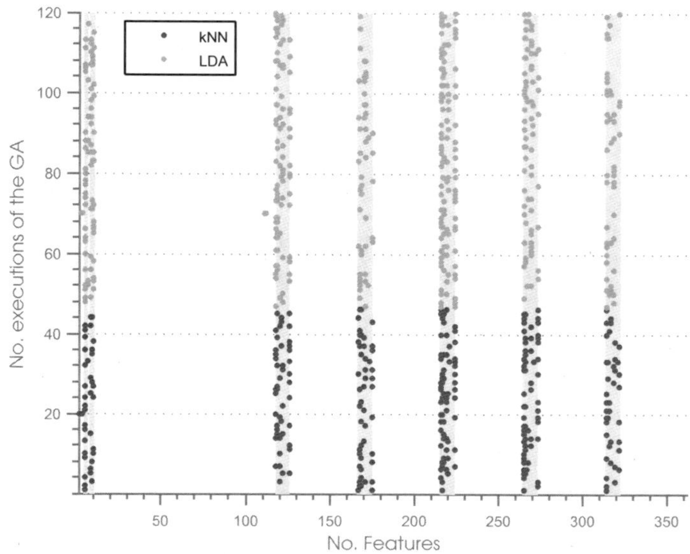

Fig. 6 Correlation between the variables of the subsets obtained. The ordenate axis indicates the number of executions of the GA. Each line corresponds to a different feature subset. The shaded regions show information about the correlated feature.

Table 3 Feature subsets which obtain the best rates in the classification phase.

No.	*Description*	*Classifier*	*Accuracy(%)*
119	Little Finger Width	kNN	99.97±0.11
126	Little Finger Aspect Proportion		
216	Middle Finger Length		
218	Middle Finger Width		
265	Index Finger Length		
266	Index Finger Width		
5	Hand Length	LDA	99.93±0.15
8	Distance Little - Index		
119	Little Finger Width		
218	Middle Finger Width		
273	Index Finger Aspect Proportion		
318	Thumb Finger Area		

they belong to the same cluster. It should be noted that the features of some clusters are more selected than others as variables in the subsets. Thus, the density of the fourth cluster, which is associated to the middle finger, is higher than the density of the last one, which is related to the thumb finger. Therefore, it could be concluded that the information obtained from the middle finger is more relevant than the thumb features. In fact, these thumb variables would not be the first choice because of its lower capability of discrimination, which is also highlighted in [26] and [29]. By

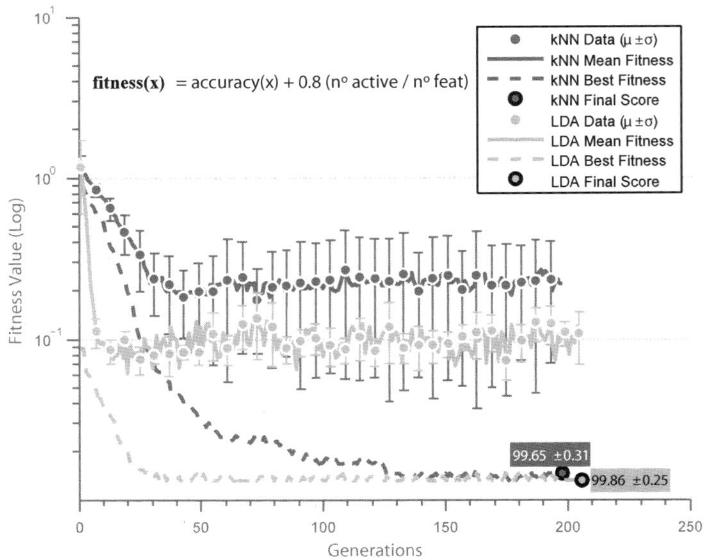

Fig. 7 Mean fitness curve of the GA performance. The lighter lines correspond to GA-LDA strategy whereas the darker ones are associated to the GA-kNN one. Dash lines represent the best accuracies in each generation.

focusing on one of the features in Figure 6, e.g. the number 216, it can be observed that it is a feature with a high probability of being selected by the GA. This implies that its significance is superior and the accuracy rates of the subsets which includes this feature will be higher than if e.g. the number 322 is the variable selected.

Thus, the Fourier descriptors and the Hu moments are not too relevant for classification tasks by using this dataset, because they have not been selected on any occasion. On the other hand, features such as length, width, area and perimeter on both the hand and the fingers are considered very discriminant and suitable because of their high rate of choice. Additionally, some parts of the hand can be considered more significant according to the number of features that they provided on average. Consequently, the middle finger together with the little and the index ones to a lesser extent, are supposed to add the most relevant information to the subset. Table 3 displays the features selected in the two subsets which provide the best accuracy rates for each method.

Figures 7 and 8 show two specific executions of the GA by applying different strategies (GA-LDA and GA-kNN). They achieve great accuracy results, almost 100%, by using just 6 features of the 362 in total. In spite of the slow convergence and high variability in each GA-kNN generation, this strategy achieves the same results in terms of accuracy as the GA-LDA one.

Table 4 compares the results of different hand classification frameworks by using the same GPDS dataset, which ensures that the results are consistent and the comparison fair. The results of our approach are considerably better than the rest of the

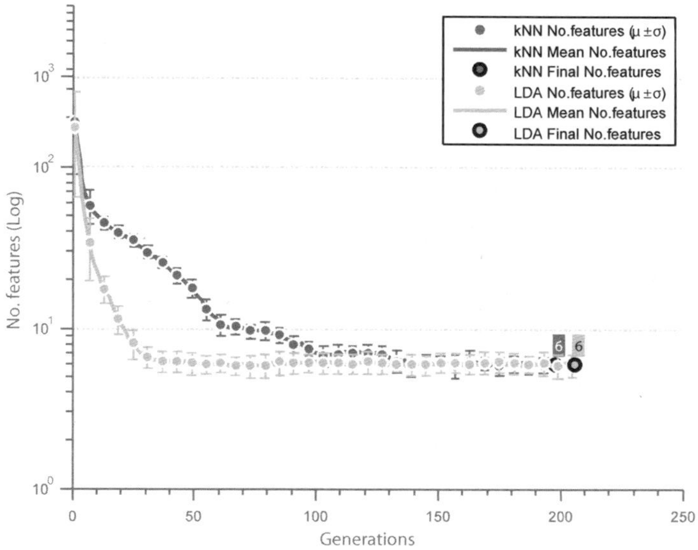

Fig. 8 Mean of the number of features in each GA generation. Two different fitness function using either LDA or kNN are applied. In the end of the performance, the number of features is similar in both methods.

Table 4 Comparison results between different hand classification systems over the same GPDS dataset.

Method	*No.Features*	*Type*	*Training(%)*	*Testing(%)*	*Accuracy(%)*
[12] MLP-ECOC BCH(31,6)	10	Geometric	30	70	98.59±0.57
			50	50	99.62±0.21
[36] Contourlet Transform - PKVA	60	Texture	30	70	98.20
Contourlet Transform - HAAR					95.60
Contourlet Transform - 9-7					96.20
[37] MLP-ECOC	15	Geometric	40	60	99.28±0.42
SVM					98.12±0.55
This GA approach - kNN	6	Geometric	30	70	**99.97±0.11**
This GA approach - LDA					**99.93±0.15**

methods, even using a smaller feature subset. Although the accuracy results in [12] (99.62%) are close to ours (99.9%),the percentage of the overall dataset that they allocate for the training set is very high with respect to ours (see Table 4). It is known that classification techniques such as Multi Layer Perceptron (MLP) and Support Vector Machines (SVM) are more utilised and complex than kNN and LDA, and they should theoretically provided better accuracy results. However, our GA strategy manages to find the most suitable features to be used in these simple classifiers, which implies that feature selection techniques can largely improve the results of the classification.

7 Conclusions

A new feature selection methodology based on genetic algorithms and mutual information and applied in hand biometric systems is presented. The aim of this proposal is to find out how important and discriminant the hand geometric features are, and if they are suitable in developing a robust and reliable biometric identification.

The number of required features have been drastically reduced from datasets with more than 350 features. In fact, high classification rates with just five or six hand features are achieved. By analysing the results, it should be noted that the feature selection task is just as, if not more important than the classification stage. The experimental results point out that simple classifiers together with our feature selection approach, get better accuracy rates (close to 100%) than other more complex classification algorithms in the literature.

The mutual information technique has complemented the results of the genetic algorithms, proving a mathematical justification to the relationship between different subset solutions. Therefore, this feature selection methodology can be applied to other biometric systems with different kinds of features, in order to improve the accuracy rate in the classification task and reduce the time complexity of the whole process.

Furthermore, a preliminary standard subset of discriminant features to be used in hand classification has been outlined. The importance of different parts of the hand has also been identified. It should be noted that all the experiments have been carried out using the GPDS dataset. In fact, a deeper analysis with more datasets should be done in further studies in order to justify the choice of a feature subset as the standard one.

Acknowledgements. This work has been partially supported by the Ministry of Science and Innovation of Spain under grant TIN2010-15351, project name 'Probabilistic self organizing models for the restoration of lossy compressed images and video', and by Junta de Andalucía (Spain) under contract TIC-01615, project name 'Intelligent Remote Sensing Systems'. The work is tested on the GPDS hand database of University of Las Palmas de Gran Canaria. Authors thankfully acknowledge the contribution made to this research due to the availability of the database.

References

1. Jain, A., Ross, A., Prabhakar, S.: An introduction to biometric recognition. IEEE Transactions on Circuits and Systems for Video Technology 14(1), 4–20 (2004)
2. Sanchez-Reillo, R., Sanchez-Avila, C., Gonzalez-Marcos, A.: Biometric identification through hand geometry measurements. IEEE Transactions on Pattern Analysis and Machine Intelligence 22(10), 1168–1171 (2000)

3. Duta, N.: A survey of biometric technology based on hand shape. Pattern Recognition 42(11), 2797–2806 (2009)
4. Fong, L.L., Seng, W.C.: A comparison study on hand recognition approaches. In: International Conference of Soft Computing and Pattern Recognition, SOCPAR, pp. 364–368 (2009)
5. Kumar, A., Wong, D., Shen, H., Jain, A.: Personal verification using palmprint and hand geometry biometric. In: Kittler, J., Nixon, M.S. (eds.) AVBPA 2003. LNCS, vol. 2688, pp. 668–678. Springer, Heidelberg (2003)
6. Zhang, D., Kong, W.K., You, J., Wong, M.: Online palmprint identification. IEEE Transactions on Pattern Analysis and Machine Intelligence 25(9), 1041–1050 (2003)
7. Esther Rani, P., Shanmuga Lakshmi, R.: Palmprint recognition system using zernike moments feature extraction. In: Das, V.V., Vijaykumar, R. (eds.) ICT 2010. CCIS, vol. 101, pp. 449–454. Springer, Heidelberg (2010)
8. Nanni, L., Lumini, A.: Ensemble of multiple palmprint representation. Expert Systems with Applications 36(3, Part 1), 4485–4490 (2009)
9. Dutağaci, H., Sankur, B., Yörük, E.: Comparative analysis of global hand appearance-based person recognition. Journal of Electronic Imaging 17(1), 011018 (2008)
10. Kumar, A., Zhang, D.: Personal recognition using hand shape and texture. IEEE Transactions on Image Processing 15(8), 2454–2461 (2006)
11. Ross, A., Jain, A.: Information fusion in biometrics. Pattern Recognition Letters 24(13), 2115–2125 (2003)
12. Faundez-Zanuy, M., Elizondo, D., Ferrer-Ballester, M.N., Travieso-Gonzàlez, C.: Authentication of individuals using hand geometry biometrics: A neural network approach. Neural Processing Letters 26, 201–216 (2007)
13. Yörük, E., Dutagaci, H., Sankur, B.: Hand biometrics. Image and Vision Computing 24(5), 483–497 (2006)
14. Connie, T., Jin, A.T.B., Ong, M.G.K., Ling, D.N.C.: An automated palmprint recognition system. Image and Vision Computing 23(5), 501–515 (2005)
15. Yang, J., Zhang, D., yu Yang, J., Niu, B.: Globally maximizing, locally minimizing: Unsupervised discriminant projection with applications to face and palm biometrics. IEEE Transactions on Pattern Analysis and Machine Intelligence 29(4), 650–664 (2007)
16. Raymer, M., Punch, W., Goodman, E., Kuhn, L., Jain, A.: Dimensionality reduction using genetic algorithms. IEEE Transactions on Evolutionary Computation 4(2), 164–171 (2000)
17. Siedlecki, W., Sklansky, J.: A note on genetic algorithms for large-scale feature selection. Pattern Recognition Letters 10(5), 335–347 (1989)
18. Yang, J., Honavar, V.: Feature subset selection using a genetic algorithm. IEEE Intelligent Systems and their Applications 13(2), 44–49 (1998)
19. Sun, Z., Bebis, G., Miller, R.: Object detection using feature subset selection. Pattern Recognition 37(11), 2165–2176 (2004)
20. McLachlan, G., Bean, R., Peel, D.: A mixture model-based approach to the clustering of microarray expression data. Bioinformatics 18(3), 413–422 (2002)
21. Cover, T., Hart, P.: Nearest neighbor pattern classification. IEEE Transactions on Information Theory 13(1), 21–27 (1967)
22. Duda, R.O., Hart, P.E., Stork, D.G.: Pattern classification, 2nd edn. Wiley (2001)
23. Guyon, I., Elisseeff, A.: An introduction to variable and feature selection. Journal of Machine Learning Research 3, 1157–1182 (2003)
24. Ferrer, M., Morales, A., Travieso, C., Alonso, J.: Low cost multimodal biometric identification system based on hand geometry, palm and finger print texture. In: 41st Annual IEEE International Carnahan Conference on Security Technology, pp. 52–58 (2007)

25. Otsu, N.: A Threshold Selection Method from Gray-level Histograms. IEEE Transactions on Systems, Man and Cybernetics 9(1), 62–66 (1979)
26. Amayeh, G., Bebis, G., Erol, A., Nicolescu, M.: Hand-based verification and identification using palm-finger segmentation and fusion. Computer Vision and Image Understanding 113(4), 477–501 (2009)
27. Mansoor, A., Mumtaz, M., Masood, H., Butt, M., Khan, S.: Personal identification using palmprint and contourlet transform. In: Bebis, G., Boyle, R., Parvin, B., Koracin, D., Remagnino, P., Porikli, F., Peters, J., Klosowski, J., Arns, L., Chun, Y.K., Rhyne, T.-M., Monroe, L. (eds.) ISVC 2008, Part II. LNCS, vol. 5359, pp. 521–530. Springer, Heidelberg (2008)
28. López-Rubio, E.: Restoration of images corrupted by Gaussian and uniform impulsive noise. Pattern Recognition 43(5), 1835–1846 (2010)
29. Oden, C., Ercil, A., Buke, B.: Combining implicit polynomials and geometric features for hand recognition. Pattern Recognition Letters 24(13), 2145–2152 (2003)
30. Barber, C., Dobkin, D., Huhdanpaa, H.: The quickhull algorithm for convex hulls. ACM Transactions on Mathematical Software (TOMS) 22(4), 469–483 (1996)
31. Liao, S.X., Pawlak, M.: On image analysis by moments. IEEE Transactions on Pattern Analysis and Machine Intelligence 18(3), 254–266 (1996)
32. Baker, J.E.: Reducing bias and inefficiency in the selection algorithm. In: Proceedings of the Second International Conference on Genetic Algorithms on Genetic Algorithms and their Application, pp. 14–21. L. Erlbaum Associates Inc (1987)
33. Guo, B., Nixon, M.: Gait feature subset selection by mutual information. IEEE Transactions on Systems, Man and Cybernetics, Part A: Systems and Humans 39(1), 36–46 (2009)
34. Leung, A., Gong, S.: Online Feature Selection Using Mutual Information for Real-time Multi-view Object Tracking. In: Zhao, W., Gong, S., Tang, X. (eds.) AMFG 2005. LNCS, vol. 3723, pp. 184–197. Springer, Heidelberg (2005)
35. López-Rubio, E.: Probabilistic self-organizing maps for qualitative data. Neural Networks 23(10), 1208–1225 (2010)
36. Mansoor, A.B., Masood, H., Mumtaz, M., Khan, S.A.: A feature level multimodal approach for palmprint identification using directional subband energies. Journal of Network and Computer Applications 34(1), 159–171 (2011)
37. Ferrer, M., Fabregas, J., Faundez, M., Alonso, J., Travieso, C.: Hand geometry identification system performance. In: 43rd Annual International Carnahan Conference on Security Technology, pp. 167–171 (2009)

Building Visual Surveillance Systems with Neural Networks

J. García-Rodríguez, A. Angelopoulou, F.J. Mora-Gimeno, and A. Psarrou

Self-organising neural networks have shown promise in a variety of applications areas. Their massive and intrinsic parallelism makes those networks suitable to solve hard problems in image-analysis and computer vision applications, especially when non-stationary environments occur. Moreover, this kind of neural networks preserves the topology of an input space by using their inherited competitive learning property. In this work we use a kind of self-organising network, the Growing Neural Gas, to solve some computer vision tasks applied to visual surveillance systems. The neural network is also modified to accelerate the learning algorithm in order to support applications with temporal constraints. This feature has been used to build a system able to track image features in video sequences. The system automatically keeps the correspondence of features among frames in the sequence using its own structure. Information obtained during the tracking process and allocated in the neural network can also be used to analyse the objects motion.

1 Introduction

Visual tracking is a very active research field in computer vision. Development of the process of visual surveillance in dynamic scenes often includes steps for modeling the environment, motion detection, classification of moving objects, tracking and recognition of actions developed. Most of the work is focused on applications related to tracking people or vehicles that have a large number of potential applications such as: controlling access to special areas, identification of people, traffic

J. García-Rodríguez · F.J. Mora-Gimeno
Dept. of Computer Technology, University of Alicante (Spain)
e-mail: {jgarcia,fjmora}@dtic.ua.es

A. Angelopoulou · A. Psarrou
Dept. of Computer Science and Software Engineering, University of Westminster (UK)
e-mail: {agelopa,psarroa}@wmin.ac.uk

D.A. Elizondo et al. (Eds.): Computational Intelligence for Privacy and Security, SCI 394, pp. 181–198.
springerlink.com

analysis, anomaly detection and management alarms or interactive monitoring using multiple cameras [1].

Some visual surveillance systems have marked important milestones. The visual surveillance system in real-time W4 [2] uses a combination of analysis of shapes and tracking, building models of appearance to detect and track groups of people and monitor their behaviors even in the presence of occlusion and in outdoor environments . This system uses a single camera and grayscale sensor. The system Pfinder [3] is used to retrieve a three-dimensional description of a person in a large space. It follows a single person without occlusions in complex scenes, and has been used in various applications. Another system to track single person, the TI [4], detects moving objects in indoor scenes using motion detection. Tracking is made using the first order prediction, and recognition is done by applying predicates to a behavior graph formed by matching objects links in successive frames. This system does not support small movements of objects in the background. The CMU system [5] can monitor activity on a wide area using multiple networked cameras. It can detect and track multiple people and vehicles in complex scenes and monitor their activities for long periods of time.

Recognition of actions has been extensively investigated [6]-[7]. The analysis of the trajectory is also one of the basic problems in understanding the actions [8]-[10].

Moreover, some visual surveillance systems for scene analysis and surveillance depend on the use of knowledge about the scenes where the objects move in a predefined manner [11]-[14].

In recent years, work in the analysis of behaviour has been successful due to the use of effective and robust techniques for detecting and tracking objects and people that have shifted the interest in higher levels of understanding of the scene. Besides the obvious advantages, the use of vision sensor networks incorporates a number of challenges such as management and fusion of data, the operation of distributed and central units of detection, and algorithmic techniques for efficient extraction of interesting behavior with efficient management redundant data. In this line, a review with recent research can be found in [15].

There are also several works that use the self-organising models for the representation and tracking of objects. Fritzke [16] proposed a variation of the Growing Neural Gas (GNG) [17] to map non–stationary distributions that [18] applies to represent and track people. The use of self-organising neural networks for human-machine interaction is suggested in [19]. In [20] amendments to self-organising models for the characterization of the movement are proposed.

From the cited works, only [18] represents both the local and the global movements. However, there is no consideration of temporal constraints, nor further exploitation of the knowledge of previous frames for segmentation and prediction in subsequent frames. Neither uses the structure of the neural network to solve the problem of correspondence in the analysis of the movement.

Considering the work in the area and previous studies about representation capabilities of self-growing neural models [21], it is proposed the design of a modular system capable of capturing images from a camera, target areas of interest and represent the morphology of entities in the scene. As well as analyzing the evolution of these entities in time and obtain semantic knowledge about the actions that occur at the scene. We propose the representation of these entities through a flexible model able to characterize their morphological and positional changes along the image sequence. The representation should identify entities over time, establishing a correspondence of these during the various observations. This feature allows the description of the behavior of the entities through the interpretation of the dynamics of the model representation.

The time constraints of the problem suggest the need for highly available systems capable to obtain representations with an acceptable quality in a limited time. Besides, the large amount of data suggests definition of parallel solutions.

To solve the problem, the representation of the objects and their motion is done with a modified self-growing model. We propose a neural architecture able to adapt the topology of the network of neurons to the shape of the entities that appear in the images, representing and characterizing objects of interest in the scenes. The model has also the ability to track the objects through a sequence of images in a robust and simple way.

In a previous work [21], it has been proved that using architectures based on GNG, the temporary restrictions on problems such as tracking objects or the recognition of gestures can be assumed. Considering the ability to manage sequences of images offering an acceptable quality of representation and refined very quickly depending on the time available.

With regard to the processing of image sequences, we have introduced several improvements to the network to accelerate the tracking and allow the architecture to work at video frequency. In this paper, the use of the GNG to represent objects in image sequences solves the costly problem of matching features over time by using the positions of neurons in the network. Likewise, the use of simple prediction facilitates the monitoring of neurons and reduces the time to readapt the network between frames without damaging the quality and speed of system response.

The data stored throughout the sequence in the structure of the neural network about characteristics of the entities represented as: position, color, texture, labels or any interesting feature, provide information on deformation, merge, paths followed by these entities and other events. This information may be analyzed and interpreted, giving the semantic description of the behaviors of these entities.

The remainder of the paper is organized as follows: section 2 provides a detailed description of the topology learning algorithm of the GNG. Section 3 presents the GNG-based tracking system to represent image sequences called GNG-Seq. Finally in section 4 some applications of the system are presented where objects appearing in images are tracked and trajectories are learned, followed by our major conclusions.

2 Topology Learning

One way to obtain a reduced and compact representation of 2D shapes is to use a topographic mapping where a low dimensional map is fitted to the high dimensional manifold of the shape, whilst preserving the topographic structure of the data. A common way to achieve this is by using self-organising neural networks where input patterns are projected onto a network of neural units such that similar patterns are projected onto units adjacent in the network and vice versa.

The approach presented in this paper is based on self-organising networks trained using the Growing Neural Gas learning method [16], an incremental training algorithm. The links between the units in the network are established through competitive hebbian learning. As a result the algorithm can be used in cases where the topological structure of the input pattern is not known a priori and yields topology preserving maps of feature manifold [22].

2.1 Growing Neural Gas

The Growing Neural Gas (GNG) [16] is an incremental neural model able to learn the topological relations of a given set of input patterns by means of competitive hebbian learning. Unlike other methods, the incremental character of this model, avoids the necessity to previously specify the network size. On the contrary, from a minimal network size, a growth process takes place, where new neurons are inserted successively using a particular type of vector quantization [23].

To determine where to insert new neurons, local error measures are gathered during the adaptation process and each new unit is inserted near the neuron which has the highest accumulated error. At each adaptation step a connection between the winner and the second-nearest neuron is created as dictated by the competitive hebbian learning algorithm. This is continued until an ending condition is fulfilled. In addition, in GNG network the learning parameters are constant in time, in contrast to other methods whose learning is based on decaying parameters. In the remaining of this Section we describe the growing neural gas algorithm.

The network is specified as:

- A set N of nodes (neurons). Each neuron $c \in N$ has its associated reference vector $w_c \in \mathcal{R}^d$. The reference vectors can be regarded as positions in the input space of their corresponding neurons.
- A set of edges (connections) between pairs of neurons. These connections are not weighted and its purpose is to define the topological structure. The edges are determined using the competitive hebbian learning algorithm. An edge aging scheme is used to remove connections that are invalid due to the activation of the neuron during the adaptation process.

1. The GNG learning algorithm is as follows:

1. Start with two neurons a and b at random positions w_a and w_b in $\mathcal{R}^d$.
2. Generate a random input signal ξ according to a density function $P(\xi)$.
3. Find the nearest neuron (winner neuron) s_1 and the second nearest s_2.

4. Increase the age of all the edges emanating from s_1.
5. Add the squared distance between the input signal and the winner neuron to a counter error of s_1:

$$\Delta error(s_1) = \left\| w_{s_1} - \xi \right\|^2 \tag{1}$$

6. Move the winner neuron s_1 and its topological neighbours (neurons connected to s_1) towards ξ by a learning step ε_w and ε_n, respectively, of the total distance:

$$\Delta w_{s_1} = \varepsilon_w(\xi - w_{s_1}) \tag{2}$$

$$\Delta w_{s_n} = \varepsilon_n(\xi - w_{s_n}) \tag{3}$$

7. If s_1 and s_2 are connected by an edge, set the age of this edge to 0. If it does not exist, create it.
8. Remove the edges larger than a_{max}. If this results in isolated neurons (without emanating edges), remove them as well.
9. Every certain number λ of input signals generated, insert a new neuron as follows:
   - Determine the neuron q with the maximum accumulated error.
   - Insert a new neuron r between q and its further neighbor f:

$$w_r = 0.5(w_q + w_f) \tag{4}$$

   - Insert new edges connecting the neuron r with neurons q and f, removing the old edge between q and f.
   - Decrease the error variables of neurons q and f multiplying them with a constant α. Initialize the error variable of r with the new value of the error variable of q and f.
10. Decrease all error variables by multiplying them with a constant β.
11. If the stopping criterion is not yet achieved, go to step 2.

In summary, the adaptation of the network to the input space takes place in step 6. The insertion of connections (step 7) between the two closest neurons to the randomly generated input patterns establishes an induced Delaunay triangulation in the input space. The elimination of connections (step 8) eliminates the edges that no longer comprise the triangulation. This is made by eliminating the connections between neurons that no longer are next or that they have nearer neurons. Finally, the accumulated error (step 5) allows the identification of those zones in the input space where it is necessary to increase the number of neurons to improve the mapping.

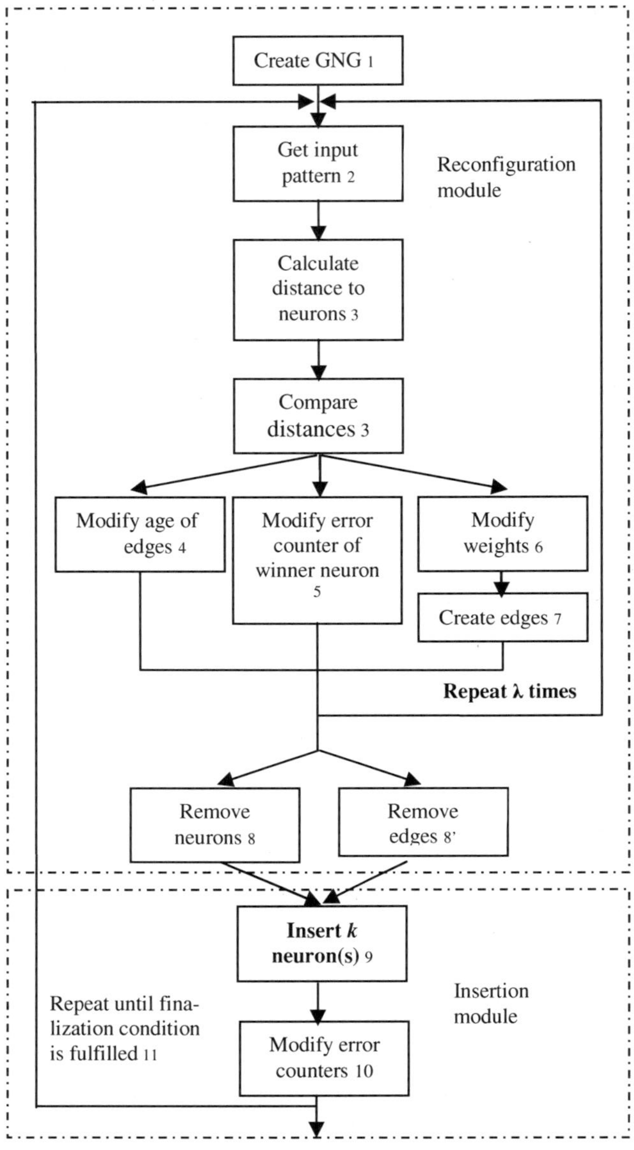

Fig. 1 GNG learning algorithm

2.2 *Accelerated Growing Neural Gas*

Computer vision and image processing tasks have in many cases temporal constraints determined by the sampling rate.

To obtain a complete network, with all its neurons, in a predetermined time, the GNG learning algorithm should be modified to accelerate its conclusion. The main factor that affects the learning time is the number λ of input signals generated by iteration (step 2), since new neurons are inserted at smaller intervals, taking less time in completing the network. Another alternative is the insertion of more than one neuron per iteration, repeating k times step 9 of the learning algorithm. In this accelerated version of the GNG, each iteration step 9 is repeated, inserting several neurons in those zones where bigger accumulated error exists, creating the corresponding connections (figure 1). This modification of GNG algorithm has interesting applications in problems with temporal restrictions. The election of the right parameters requires a previous experimentation with the problem to be solved [21].

Fig. 2 Representation of two-dimensional objects with a self-organising network.

3 Image Sequences Representation with GNG

Previous step to track entities in image sequences is the adaptation of the GNG algorithm to use images as input space. This way, reference vectors of GNG neurons will adapt and represent objects appearing in images.

3.1 Representation of 2D Objects with GNG

Given an image $I(x,y) \in \mathcal{R}$ we perform the transformation $\psi_{\mathcal{T}}(x,y) = \mathcal{T}(I(x,y))$ that associates to each one of the pixels its probability of belonging to the object, according to a property $\mathcal{T}$, for instance a threshold function.

If we consider $\xi = (x,y)$ and $\mathcal{P}(\xi) = \psi_{\mathcal{T}}(\xi)$, we can apply the learning algorithm of the GNG to the image I, so that the network adapts its topology to the object shape. This adaptive process is iterative, so the GNG represents the object during all the learning.

As a result of the GNG learning process we obtain a graph that we call Topology Preserving Graph $\mathcal{TPG} = \langle \mathcal{N}, C \rangle$, with a vertex (neurons) set $\mathcal{N}$ and an edge set C that connect them (figure 2). This *TPG* establishes a Delaunay triangulation induced by the object.

3.2 System Overview

The technique used to represent the objects shape and analyze its movement is based on the tracking throughout frames of some features of the object representation obtained with the neural network structure. The reference vectors of the neural network neurons are used like markers to follow, since no neurons are added or deleted.

One of the most advantageous characteristics of the graph obtained for each of the objects that appear in any image frame from the sequence is that no reset of the learning of the network is required. Besides, the representation obtained for the previous frame can be used, whenever the speed of sampling is high. We have called this version GNG-Seq.

The tracking process would be based on the following scheme: (figure 3)

- Calculation of the transformation function ψ in order to segment the objects from the background based on information from previous maps.
- Prediction of the new positions of the neurons reference vector.
- Re-adjustment of the neurons reference vector.

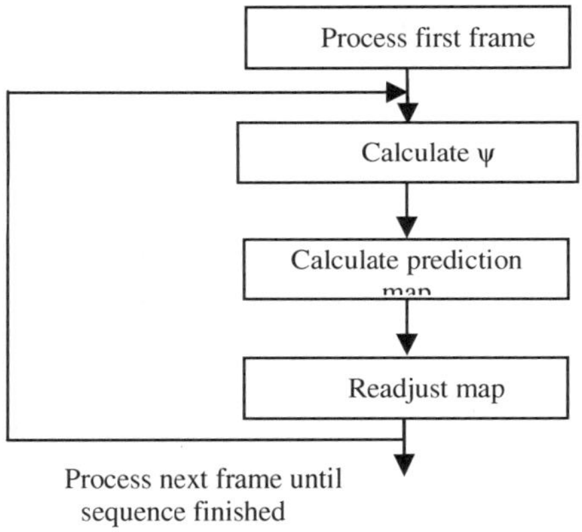

Fig. 3 GNG-Seq system

The main difference with the accelerated GNG algorithm is the omission of insertion/deletion actions (steps 8 to 11) after the first frame. Since no neurons are added or deleted the system keeps correspondence during the whole sequence, solving intrinsically this problem. Following the dynamic of the neural network structure and using the neurons to predicting and readjusting the representation. For the initial moment t_0 the representation is obtained making a complete adaptation of a GNG. However, for the following frames the previous network structure is employed. So, the new representation of the object is obtained by performing the iteration of the internal loop of the learning algorithm of the GNG, relocating the neurons and creating or removing edges. This adaptive method is also able to face real-time constraints, because the number λ of times that the internal loop is performed can be chosen according to the available time between two successive frames that depends on the video rate.

3.3 Segmentation

Extraction of objects from sequences of images makes it necessary to update the transformation function for each one of the images since the conditions of illumination, noise, etc can change even between consecutive frames. For surveillance applications neurons reference vectors could be used to restrict the search area.

Figure 4 shows the scheme used to distinguish the treatment of the first frame and the later in which information is available on the maps and learn from previous frames in the sequence.

Not only the transformation function is calculated but, the system can update the values of property T that determine what the object is at the moment of segmentation. Therefore points of the object from previous frames are known because we have built the correspondent *TPG*. That is:

$$\psi_T(x, y, t) = T(I(x, y, t), TPG^{t-n} \quad (5)$$

Where n can represent the number of previous frames used to calculate a prediction of segmentation, representation and position of objects in the current frame.

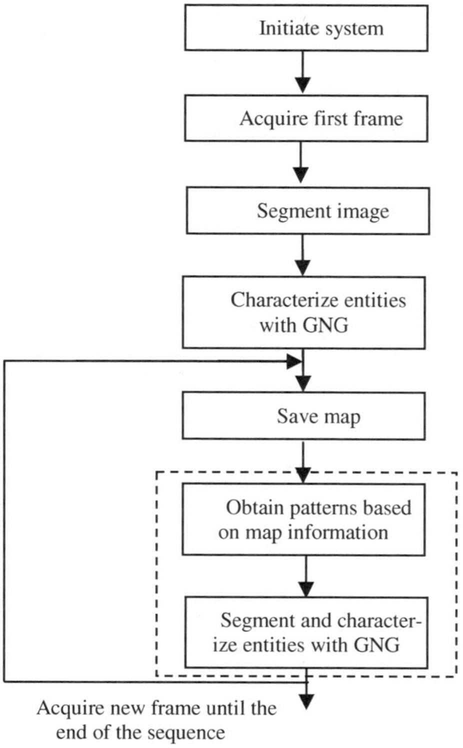

Fig. 4 GNG-seq algorithm remarking prediction applied to segmentation.

In visual surveillance applications, frame differencing or background modelling techniques need to be used to segment objects of interest. In this case information of neurons reference vectors should be used to draw a square around the objects of interest that defines the area where objects should appear in next frames. In this way, it is only necessary to segment inside these squares and only every number of frames would it be necessary to update the model to support new objects that appear in the scene (figure 5). Considering that all the images have been acquired with static cameras.

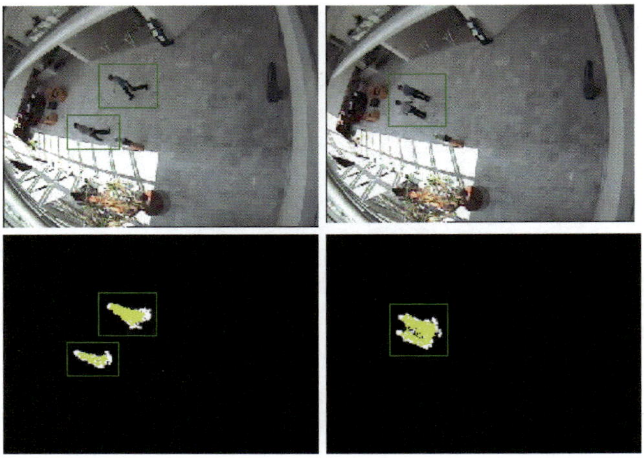

Fig. 5 Object segmentation based on GNG.

3.4 Prediction

Calculation of motion and prediction of new locations in many tracking systems are based on probabilistic methods (Kalman filters [24], particle filters or CONDENSATION algorithm [25]). The system requires an accurate model of motion and noise. These models are computationally expensive when some objects or different parts of the same object need to be tracked. We employ a simple approach based on neurons trajectories which allow motion vector computation based on positions of the neurons at time *t-1* and *t-2*. This way, velocity and acceleration of the objects or parts of objects tracked are obtained.

Prediction allows the system to segment the object in the new image, to predict the new position and to readjust the map based on the information from previous maps. Since the speed and the acceleration of the different nodes of the graph, we can predict the new position of the *TPG* before iterating the internal loop.

As example of the importance of implementing a prediction system, figure 6 shows the adaptation of the GNG to a new frame. At the top is shown the starting frame. On the left, prediction was applied and the evolution is good. However, on the right image prediction has not been applied and the adjustment is incorrect.

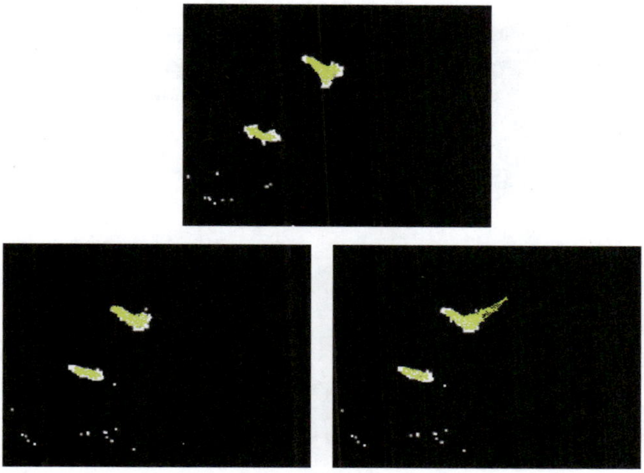

Fig. 6 Examples of adaptation with and without prediction

3.5 Common and Relative Motion

Analysis of the common motion can be made following through the frame sequence the centroid of the object, based on the centroid of the neurons reference vectors that represent it. In this way, a single trajectory is defined for the object. Figure 7 (central row) shows an example of common motion representation of two people appearing in a video sequence.

That is, the common movement is regarded as the trajectory described by the centroid of the *TPG* obtained with the GNG:

$$M_C = Traj_{c_m} = \{c_{m_{t0}}, \dots, c_{m_{tf}}\} \quad (6)$$

The relative movement of an object is interpreted like the trajectories followed by each one of the nodes of the graph created during the learning algorithm, the *TPG*:

$$M = [Traj_i], \forall i \in A \quad (7)$$

where the trajectory comes determined by the succession of positions (reference vectors) for each one of the neurons in the different frames of the image sequence:

$$Traj_i = \{w_{i_{t_0}}, \dots, w_{i_{t_f}}\} \quad (8)$$

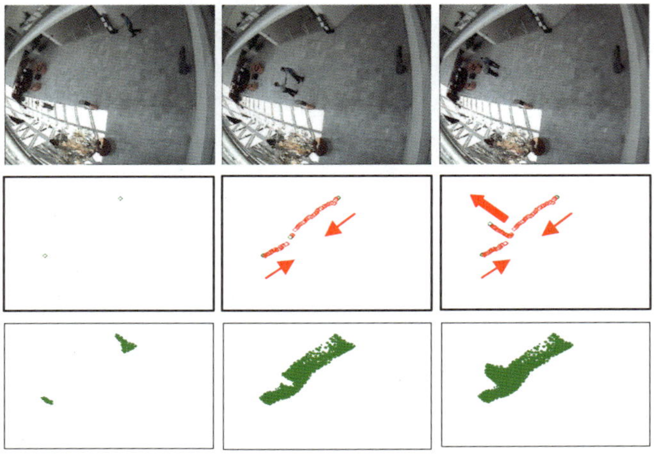

Fig. 7 Representation of common and relative motion of a sequence from the CAVIAR database.

Figure 7 shows the original frames on the upper row, common movement on the middle and relative movement on the bottom.

3.6 *Tracking Multiple Objects*

There are several studies on the labelling and tracking of multiple objects. Some of them based on the trajectory [26] or the current state [27]-[28] explore the way in which they interact. There is also an important field of study in related problems such as occlusions [29].

The technique used here for tracking multiple objects is based on the GNG-Seq which using its fast algorithm separates the different objects in the image. Once the objects in the image are separated, it is possible to identify groups of neurons that are mapping each object and follow them separately. These groups will be identified and labelled to use them as a reference and keep the correspondence between frames. In practice, the method is implemented through a recursive algorithm to traverse all neurons, followed by the structures of the neighborhood map of the connections that define each group.

3.6.1 Merger and Division

The ability of growing neural gas to divide its structure to map all the input space is especially useful for objects that are divided or multiple objects in the image. The network will eliminate unnecessary edges so that objects are represented independently by groups of neurons.

If the input space is unified again, the network will adapt these changes by introducing new edges that reflect homogeneous input spaces. Since neurons are not added or deleted, the objects or persons that come together and split into groups, can be identify even track them separately or together.

This last feature is a great advantage of the representation model that gives the system great versatility in terms of track entities or groups of entities in video sequences.

The merge of entities is represented as the union of the Topology Preserving Graph that mapped entities.

$$GPT_1 \cup GPT_2 \cup \ldots \cup GPT_n \Rightarrow GPT_G \tag{9}$$

In the case of division of entities, the map that represents the group splits in different clusters.

$$GPT_G \Rightarrow (GPT_1, GPT_2, \ldots, GPT_n) \tag{10}$$

3.6.2 Occlusions

The modelling of individual objects does not consider the interaction between multiple objects or the interaction of these objects with the background during the tracking. For example, in the case of partial or total occlusions occur between objects.

The way in which the occlusions are handled in this work is to discard the image if the object is completely concealed by the background. In each image, once an object is segmented and characterized, the number of pixels that represent each object in the original image is calculated. Frames are discarded, if percentage of pixels loss with respect to the average value calculated for the previous frames is very high and resumed the consideration of frames when the rate becomes again acceptable. In the case of partial occlusion with the background or between objects it would be expected to adapt the representation after a few frames.

4 Experiments

Figure 8 shows the representation of three frames of an image sequence from CAVIAR database project [30] where two people walk in a lobby.

The system tracks the people and represents the trajectories based on the representation obtained from the GNG. The model is able to separate in two clusters people walking alone and to join the two clusters into a single one when people walk together in the lobby. Images are processed at a video rate speed since the re-adjustment for frames after the first one requires only one loop in the learning algorithm.

The moving objects are detected using a temporal differencing approach. The search area is restricted by using a square defined by the positions of the different clusters obtained with the application of GNG to the image. A prediction is also applied that moves the neurons of the map calculating the motion vector based on the position (reference vectors) of the neurons in previous frames.

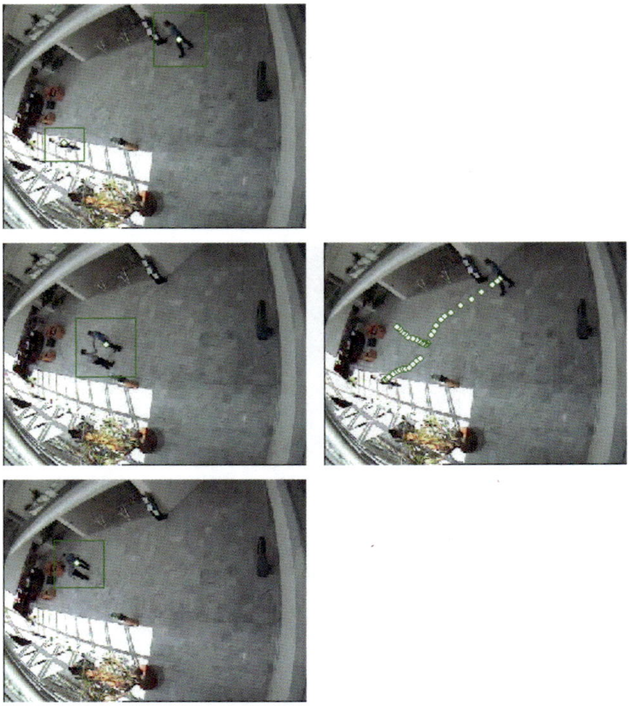

Fig. 8 Representation of some frames of a image sequence from the CAVIAR database project tracking people and representing trajectories.

Figures 9 and 10 represent sequences from the CAVIAR database where people joint and walk together or separate in different groups. On the right column, original frames are represented. On the middle, the GNG representation onto segmented images is shown and on the right column, the trajectories of the clusters obtained from the GNG representation are presented. The different rows represent firs frame (top), middle frame (middle) and last frame (bottom) of each sequence.

The common motion would it be extracted from GNG structure represented in central column of figures 9 and 10. It should be used to determine abnormal behaviours of individual people, evaluating deformations of GNG representation to determine behaviours of people like: walk, run, fight or many others.

Relative motion represented on the right column of figures 9 and 10 provides information about trajectories followed by people appearing in the image sequence. This information permits the study of global actions developed in the scene.

The computational cost of the GNG representation system is very low, since after first frame no neurons are added or deleted and only the reconfiguration module of figure 1 need to be executed. The system is able to work at video rate.

The capacity of the system to manage local and global movement improves the optical flow based systems performance.

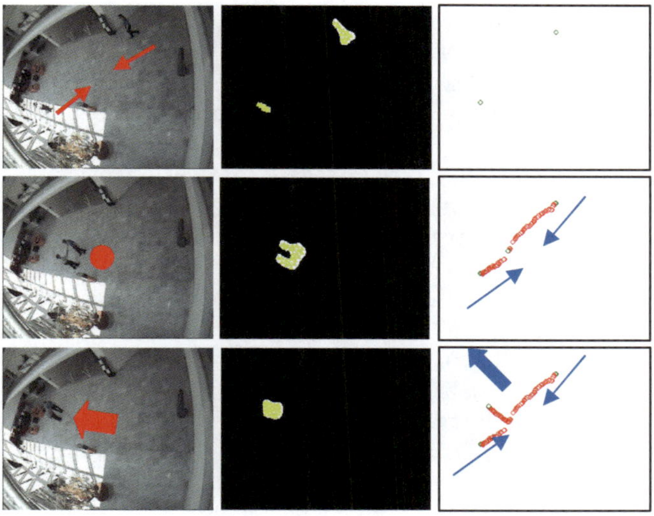

Fig. 9 Merge sequence representation.

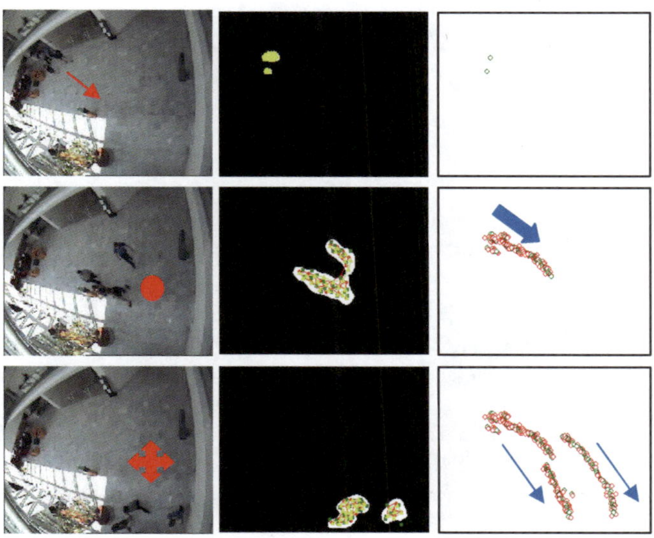

Fig. 10 Split sequence representation.

5 Conclusions and Further Work

In this work we present a surveillance system based on GNG neural network capable of representing motion under time constraints.

The proposed system incorporates mechanisms for prediction based on information stored within the network structure on the characteristics of objects such as color, shape or situation to anticipate certain operations such as segmentation and positioning of objects in subsequent frames. This provides for a more rapid adaptation to the objects in the image, restricting the areas of search and anticipating the new positions of objects.

Processing information on the neurons reference vectors of the map along the time is possible to construct the path followed by objects, represent and analyse the sequence. This evolution can be studied from global movement, using the centroids of the map or from local movement, by studying the deformations of the object. The latter is possible because the system does not reset the map for each image of the sequence. It only readjusts the network structure starting from previous map without inserting or deleting neurons. In this way the neurons are used as markers that define the stable form of objects.

The system automatically handles the mergers and divisions among entities that appear in the images and can detect and interpret the actions that are performed in video sequences.

Improvements in segmentation and prediction methods will be included in further versions. Performance evaluation of the model with other databases will be provided, testing the system under more challenging sequences. The system architecture is also being implemented with NVDIA CUDA C to exploit the parallel features of images and neural networks.

Acknowledgments. This work was partially supported by the University of Alicante project GRE09-16 and Valencian Government project GV/2011/034.

References

[1] Hu, W., Tan, T., Wang, L., Maybank, S.: A Survey on Visual Surveillance of Object Mo-tion Behaviors. IEEE Transactions on Systems, Man and Cybernetics 34(3), 334–352 (2004)

[2] Haritaoglu, H., Harwood, D., Davis, L.S.: W4: Who? When? Where? What? A Real Time System for Detecting and Tracking People. In: Proceedings of the International Conference on Automatic Face and Gesture Recognition, pp. 222–227 (1998)

[3] Wren, C., Azarbayejani, A., Darell, T., Pentland, A.: Pfinder: Real-Time Tracking of the Human Body. IEEE Transactions on Pattern Analysis and Machine Intelligence 19(7), 780–785 (1997)

[4] Olson, T., Brill, F.: Moving Object Detection and Event Recognition Algorithms for Smart Cameras. In: Proc. DARPA Image Understanding Workshop, pp. 159–175 (1997)

[5] Lipton, A.J., Fujioshi, H., Patil, R.S.: Moving Target Classification and Tracking from Real-Time Video. In: Proc. IEEE Workshop Applications of Computer Vision, pp. 8–14 (1998)

[6] Collins, T., Lipton, A.J., Kanade, T.: Introduction to the special section on video surveillance. IEEE Trans. Pattern Anal. Machine Intell. 22, 745–746 (2000)

[7] Howarth, R.J., Buxton, H.: Conceptual descriptions from monitoring and watching im-age sequences. Image and Vision Computing 18(9), 105–135 (2000)

[8] Hu, W., Xie, D., Nan, T.N.: A Hierarchical Self-Organizing Approach for Learning the Patterns of Motion Trajectories. Chin. J. Comput. 26(4), 417–426 (2003)

[9] Buxton, H.: Learning and Understanding dynamic scene activity: a review. Image. Vis. Comput. 21(1), 125–136 (2003)

[10] Bremond, F., Thonnat, M., Zuñiga, M.: Video understanding framework for automat-ic behaviour recognition. Behav. Res. Meth. (2006)

[11] Brand, M., Kettnaker, V.: Discovery and Segmentation of Activities in Video. IEEE Trans. Pattern Anal. Mach. Intell. 22(8), 844–885 (2000)

[12] Makris, D., Ellis, T.: Learning semantic scene models from observating activity in vis-ual surveillance. IEEE Trans. Syst. Man Cybern. 35(3), 397–408 (2005)

[13] Mckenna, S.J., Nait Charif, H.: Summarising contextual activity and detecting unsual inac-tivity in a supportive home environment. Pattern Anal. Appl. 7(4), 386–401 (2004)

[14] Stauffer, C.: Estimating tracking sources and sinks. In: Proceedings of 2nd IEEE workshop of event mining, pp. 259–266 (2003)

[15] Dee, H.M., Velastin, S.: How close are we to solving the problem of automated visual surveillance? Machine Vision and Applications 19, 329–343 (2008)

[16] Fritzke, B.: A Self-Organizing Network that Can Follow Non-stationary Distributions. In: Proceedings of the 7th International Conference on Artificial Neural Networks, pp. 613–618 (1997)

[17] Fritzke, B.: A Growing Neural Gas Network Learns Topologies. In: Tesauro, G., Touretzky, D.S., Leen, T.K. (eds.) Advances in Neural Information Processing Systems, vol. 7, pp. 625–632. MIT Press (1995)

[18] Frezza-Buet, H.: Following Non-stationary Distributions by Controlling the Vector Quatization Accuracy of a Growing Neural Gas Network. Neurocomputing 71, 1191–1202 (2008)

[19] Stergiopoulou, E., Papamarkos, N., Atsalakis, A.: Hand Gesture Recognition Via a New Self-Organized Neural Network. In: Sanfeliu, A., Cortés, M.L. (eds.) CIARP 2005. LNCS, vol. 3773, pp. 891–904. Springer, Heidelberg (2005)

[20] Cao, X., Suganthan, P.N.: Video Shot Motion Characterization Based on Hierachical Overlapped Growing Neural Gas Networks. Multimedia Systems 9, 378–385 (2003)

[21] García, J., Flórez-Revuelta, F., García, J.M.: Growing Neural Gas for Vision Tasks with Time Restrictions. In: Kollias, S.D., Stafylopatis, A., Duch, W., Oja, E. (eds.) ICANN 2006. LNCS, vol. 4132, pp. 578–586. Springer, Heidelberg (2006)

[22] Martinetz, T., Shulten, K.: Topology Representing Networks. Neural Networks 7(3), 507–522 (1994)

[23] Kohonen, T.: Self-Organizing Maps. Springer, Heidelberg (2001)

[24] Welch, G., Bishop, G.: An Introduction to the Kalman Filter. In: ACM SIGGRAPH, Course 8 (2001), `http://www.cs.unc.edu/~welch/kalman/,2001`

[25] Ristic, B., Arulampalam, S., Gordon, N.: Beyond the Kalman Filter. Particle Filters for Tracking Applications. Artech House (2004)

[26] Han, M., Xu, W., Tao, H., Gong, Y.: An Algorithm for Multiple Object Trajectory Tracking. In: Proceedings of IEEE Computer Vision and Pattern Recognition Conference (2004)

[27] Hue, C., Le Cadre, J.C., Pérez, P.: Tracking Multiple Objects with Particle Filtering. IEEE Transactions on Aerospace and Electronic Systems 38(3), 791–812 (2002)
[28] Sullivan, J., Carlsson, S.: Tracking and Labelling of Interacting Multiple Targets. In: Leonardis, A., Bischof, H., Pinz, A. (eds.) ECCV 2006. LNCS, vol. 3953, pp. 619–632. Springer, Heidelberg (2006)
[29] Huang, Y., Essa, I.: Tracking Multiple Objects trough Occlusions. In: Proceedings IEEE Conference on Computer Vision and Pattern Recognition (2005)
[30] Fisher, R.B.: PETS 2004 Surveillance Ground Truth Data Set. In: Proc. Sixth IEEE Int. Work. on Performance Evaluation of Tracking and Surveillance (PETS 2004), pp. 1–5 (2004)

Advances in Spam Filtering Techniques

Tiago A. Almeida and Akebo Yamakami

Abstract. Nowadays e-mail spam is not a novelty, but it is still an important rising problem with a big economic impact in society. Fortunately, there are different approaches able to automatically detect and remove most of those messages, and the best-known ones are based on machine learning techniques, such as Naïve Bayes classifiers and Support Vector Machines. However, there are several different models of Naïve Bayes filters, something the spam literature does not always acknowledge. In this chapter, we present and compare seven different versions of Naïve Bayes classifiers, the well-known linear Support Vector Machine and a new method based on the Minimum Description Length principle. Furthermore, we have conducted an empirical experiment on six public and real non-encoded datasets. The results indicate that the proposed filter is easy to implement, incrementally updateable and clearly outperforms the state-of-the-art spam filters.

1 Introduction

E-mail is one of the most popular, fastest and cheapest means of communication which has become a part of everyday life for millions of people, changing the way we work and collaborate. The downside of such a success is the constantly growing volume of e-mail spam we receive.

The term *spam* is generally used to denote an unsolicited commercial e-mail. Spam messages are annoying to most users because they clutter their mailboxes. It can be quantified in economical terms since many hours are wasted everyday by workers. It is not just the time they waste reading the spam but also the time they spend removing those messages.

Tiago A. Almeida · Akebo Yamakami
School of Electrical and Computer Engineering,
University of Campinas – UNICAMP, 13081-970, Campinas, Sao Paulo, Brazil
e-mail: `{tiago,akebo}@dt.fee.unicamp.br`

D.A. Elizondo et al. (Eds.): Computational Intelligence for Privacy and Security, SCI 394, pp. 199–214.
springerlink.com

The amount of spam is frightfully increasing. The average of spams sent per day increased from 2.4 billion in 2002[1] to 300 billion in 2010[2] representing more than 90% of all incoming e-mail. On a worldwide basis, the total cost in dealing with spam was estimated to rise from US$ 20.5 billion in 2003, to US$ 198 billion in 2009.

Many methods have been proposed to automatically classify messages as spams or legitimates. Among all techniques, machine learning algorithms have achieved more success [9]. These methods include approaches that are considered top-performers in text categorization, like support vector machines and Naïve Bayes classifiers.

A relatively recent method for inductive inference which is still rarely employed in text categorization tasks is the Minimum Description Length principle. It states that the best explanation, given a limited set of observed data, is the one that yields the greatest compression of the data [8, 12, 23].

In this chapter, we present a spam filtering approach based on the Minimum Description Length principle and compare its performance with seven different models of Naïve Bayes classifiers and the linear Support Vector Machine. Here, we carry out an evaluation with the practical purpose of filtering e-mail spams in order to review and compare the currently top-performers spam filters. We have conducted an empirical experiment using six well-known, large, and public databases and the reported results indicate that our approach outperforms currently established spam filters.

Separated pieces of this chapter were presented at IEEE ICMLA 2009 [2], ACM SAC 2010 [3, 4], and IEEE IJCNN 2010 [1]. Here, we have connected all ideas in a very consistent way. We have also offered a lot more details about each study and significantly extended the performance evaluation.

The remainder of this chapter is organized as follows. Section 2 presents the basic concepts regarding the main spam filtering techniques. In Section 3, we describe a new approach based on the Minimum Description Length principle. Section 4 presents details of the Naïve Bayes algorithms applied in spam filtering domain. The linear Support Vector Machine classifier is described in Section 5. Experimental results are shown in Section 6. Finally, Section 7 offers conclusions and outlines for future work.

2 Basic Concepts

(Acc_w% In general, the machine learning algorithms applied to spam filtering can be summarized as follows.

Given a set of messages $\mathcal{M} = \{m_1, m_2, \ldots, m_j, \ldots, m_{|\mathcal{M}|}\}$ and a category set $\mathcal{C} = \{\text{spam } (c_s), \text{ legitimate } (c_l)\}$, where m_j is the jth mail in $\mathcal{M}$ and $\mathcal{C}$ is the possible label set, the task of automated spam filtering consists in building a Boolean

1 See `http://www.spamlaws.com/spam-stats.html`

2 See `www.ciscosystems.cd/en/US/prod/collateral/cisco_2009_asr.pdf`

categorization function $\Omega(m_j, c_i) : \mathscr{M} \times \mathscr{C} \rightarrow \{\texttt{True}, \texttt{False}\}$. When $\Omega(m_j, c_i)$ is `True`, it indicates message m_j belongs to category c_i; otherwise, m_j does not belong to c_i.

In the setting of spam filtering there exist only two category labels: `spam` and `legitimate` (also called `ham`). Each message $m_j \in \mathscr{M}$ can only be assigned to one of them, but not to both. Therefore, we can use a simplified categorization function $\Omega_{\texttt{spam}}(m_j) : \mathscr{M} \rightarrow \{\texttt{True}, \texttt{False}\}$. Hence, a message is classified as spam when $\Omega_{\texttt{spam}}(m_j)$ is `True`, and legitimate otherwise.

The application of supervised machine learning algorithms for spam filtering consists of two stages:

1. *Training*. A set of labeled messages ($\mathscr{M}$) must be provided as training data, which are first transformed into a representation that can be understood by the learning algorithms. The most commonly used representation for spam filtering is the vector space model, in which each document $m_j \in \mathscr{M}$ is transformed into a real vector $\mathbf{x}_j \in \Re^{|\Phi|}$, where Φ is the vocabulary (feature set), and the coordinates of $\mathbf{x}_j$ represent the weight of each feature in Φ. Then, we can run a learning algorithm over the training data to create a classifier $\Omega_{\texttt{spam}}(\mathbf{x_j}) \rightarrow \{\texttt{True}, \texttt{False}\}$.
2. *Classification*. The classifier $\Omega_{\texttt{spam}}(\mathbf{x_j})$ is applied to the vector representation of a message $\mathbf{x}$ to produce a prediction whether $\mathbf{x}$ is spam or not.

3 Spam Filter Based on Data Compression Model

The Minimum Description Length (MDL) principle is a formalization of Occam's Razor in which the best hypothesis for a given set of data is the one that yields compact representations. The traditional MDL principle states that the preferred model results in the shortest description of the model and the data, given this model. In other words, the model that best compresses the data is selected. This model selection criterion naturally balances the complexity of the model and the degree to which this model fits the data. This principle was first introduced by Rissanen [23] and it becomes an important concept in information theory.

The main concept behind the MDL principle can be presented as following. Let $\mathscr{Z}$ be a finite or countable set and let P be a probability distribution on $\mathscr{Z}$. Then there exists a prefix code C for $\mathscr{Z}$ such that for all $z \in \mathscr{Z}$, $L_C(z) = \lceil -log_2 P(z) \rceil$. C is called the code corresponding to P. Similarly, let C be a prefix code for $\mathscr{Z}$. Then there exists a (possibly defective) probability distribution P such that for all $z \in \mathscr{Z}$, $-log_2 P'(z) = L_{C'}(z)$. P' is called the probability distribution corresponding to C'. Thus, large probability according to P means small code length according to the code corresponding to P and vice versa [8, 12, 23].

The goal of statistical inference may be cast as trying to find regularity in the data. Regularity may be identified with ability to compress. MDL combines these two insights by viewing learning as data compression: it tells us that, for a given set of hypotheses $\mathscr{H}$ and data set $\mathscr{D}$, we should try to find the hypothesis or combination of hypotheses in $\mathscr{H}$ that compresses $\mathscr{D}$ most [8, 12, 23].

In essence, compression algorithms can be applied to text categorization by building one compression model from the training documents of each class and using these models to evaluate the target document.

In this way, given a set of pre-classified training messages $\mathcal{M}$, the task is to assign a target e-mail m with an unknown label to one of the classes $c \in \{spam, ham\}$. So, the method measures the increase of the description length of the data set as a result of the addition of the target document. Finally, it chooses the class for which the description length increase is minimal.

Assuming that each class (model) c is a sequence of terms extracted from the messages and inserted into the training set, each term (token) t from m has a code length L_t based on the sequence of terms presented in the messages of the training set of c. The length of m when assigned to the class c corresponds to the sum of all code lengths associated with each term of m, $Lm = \sum_{i=1}^{|m|} L_{t_i}$. We calculate $L_{t_i} = \lceil -log_2 P_{t_i} \rceil$, where P is a probability distribution related with the terms of class. Let $n_c(t_i)$ the number of times that t_i appears in messages of class c, then the probability that any term belongs to c is given by the maximum likelihood estimation:

$$P_{t_i} = \frac{n_c(t_i) + \frac{1}{|\Phi|}}{n_c + 1}$$

where n_c corresponds to the sum of $n_c(t_i)$ for all terms that appear in messages that belongs to c and $|\Phi|$ is the vocabulary size. In this chapter, we assume that $|\Phi| = 2^{32}$, that is, each term in an uncompress mode is a symbol with 32 bits. This estimation reserves a "portion" of probability to words which the classifier has never seen before.

Basically, the MDL spam filter classify a message by following these steps:

1. Tokenization: the classifier extract all terms of the new message $m = \{t_1, \ldots, t_{|m|}\}$;
2. Compute the increase of the description length when m is assigned to each class $c \in \{spam, ham\}$:

$$L_m(spam) = \sum_{i=1}^{|m|} \left\lceil -log_2 \left(\frac{n_{spam}(t_i) + \frac{1}{|\Phi|}}{n_{spam} + 1} \right) \right\rceil$$

$$L_m(ham) = \sum_{i=1}^{|m|} \left\lceil -log_2 \left(\frac{n_{ham}(t_i) + \frac{1}{|\Phi|}}{n_{ham} + 1} \right) \right\rceil$$

3. if $L_m(spam) > L_m(ham)$, then m is classified as spam; otherwise, m is labeled as ham.
4. Training method.

In the following, we offer more details about the steps 1 and 4.

3.1 Preprocessing and Tokenization

We did not perform language-specific preprocessing techniques such as word stemming, stop word removal, or case folding, since other works found that such techniques tend to hurt spam-filtering accuracy [9, 22, 32].

Tokenization is the first stage in the classification pipeline; it involves breaking the text stream into terms ("words"), usually by means of a regular expression. We consider in this chapter that terms start with a printable character; followed by any number of alphanumeric characters, excluding dots, commas and colons from the middle of the pattern. With this pattern, domain names and mail addresses will be split at dots, so the classifier can recognize a domain even if subdomains vary [28]. As proposed by Drucker *et al.* [10], and Metsis *et al.* [22], we do not consider the number of times a term appears in each message. In this way, each term is computed only once per message it appears.

3.2 Training Method

Spam filters generally build their predicting models by learning from examples. A basic training method is to start with an empty model, classify each new sample and train it in the right class if the classification is wrong. This is known as train on error (TOE). An improvement to this method is to train also when the classification is right, but the score is near the boundary that is, train on or near error (TONE). This method is also called thick threshold training [28].

The advantage of TONE over TOE is that it accelerates the learning process by exposing the filter to additional hard-to-classify samples in the same training period. Therefore, we employ the TONE as training method used by the proposed MDL anti-spam filter.

A good point of the MDL classifier is that we can start with an empty training set and according to the user feedback the classifier builds the models for each class. Moreover, it is not necessary to keep the messages used for training since the models are incrementally building by the term frequencies.

4 Naïve Bayes Spam Filters

Probabilistic classifiers are historically the first proposed filters. From Bayes' theorem and the theorem of the total probability, the probability for a message with vector $\mathbf{x} = \langle x_1, \ldots, x_n \rangle$ belongs to a category $c_i \in \{c_s, c_l\}$ is:

$$P(c_i|\mathbf{x}) = \frac{P(c_i).P(\mathbf{x}|c_i)}{P(\mathbf{x})}.$$

Since the denominator does not depend on the category, Naïve Bayes (NB) filter classifies each message in the category that maximizes $P(c_i).P(\mathbf{x}|c_i)$. In the spam filtering domain it is equivalent to classify a message as spam (c_s) whenever

$$\frac{P(c_s).P(\mathbf{x}|c_s)}{P(c_s).P(\mathbf{x}|c_s)+P(c_l).P(\mathbf{x}|c_l)} > T,$$

with $T = 0.5$. By varying T, we can opt for more true negatives (legitimate messages correctly classified) at the expense of fewer true positives (spam messages correctly classified), or vice-versa. The *a priori* probabilities $P(c_i)$ can be estimated as occurrences frequency of documents belonging to the category c_i in the training set $\mathscr{M}$, whereas $P(\mathbf{x}|c_i)$ is practically impossible to estimate directly because we would need in $\mathscr{M}$ some messages identical to the one we want to classify. However, the NB classifier makes a simple assumption that the terms in a message are conditionally independent and the order they appear is irrelevant. The probabilities $P(\mathbf{x}|c_i)$ are estimated differently in each NB model.

Despite the fact that its independence assumption is usually oversimplistic, several studies have found the NB classifier to be surprisingly effective in the spam filtering task [5, 6, 19].

The NB classifiers are the most employed in proprietary and open-source systems proposed for spam filtering [19, 22, 29]. However, there are different models of Naïve Bayes filters, something the spam literature does not always acknowledge.

In the following, we describe seven different models of NB spam filter available in the literature.

4.1 Basic Naïve Bayes

We call Basic NB the first NB spam filter proposed by Sahami *et al.* [24]. Let $\Phi = \{t_1,\ldots,t_n\}$ the set of terms, each message m is represented as a binary vector $\mathbf{x} = \langle x_1,\ldots,x_n \rangle$, where each x_k shows whether or not t_k will occur in m. The probabilities $P(\mathbf{x}|c_i)$ are calculated by:

$$P(\mathbf{x}|c_i) = \prod_{k=1}^{n} P(t_k|c_i),$$

and the criterion for classifying a message as spam is:

$$\frac{P(c_s).\prod_{k=1}^{n} P(t_k|c_s)}{\sum_{c_i\in\{c_s,c_l\}} P(c_i).\prod_{k=1}^{n} P(t_k|c_i)} > T.$$

Here, probabilities $P(t_k|c_i)$ are estimated by:

$$P(t_k|c_i) = \frac{|\mathscr{M}_{t_k,c_i}|}{|\mathscr{M}_{c_i}|},$$

where $|\mathscr{M}_{t_k,c_i}|$ is the number of training messages of category c_i that contain the term t_k, and $|\mathscr{M}_{c_i}|$ is the total number of training messages that belong to the category c_i.

4.2 Multinomial Term Frequency Naïve Bayes

The multinomial term frequency NB (MN TF NB) represents each message as a set of terms $m = \{t_1, \ldots, t_n\}$, computing each one of t_k as how many times it appears in m. In this sense, m can be represented by a vector $\mathbf{x} = \langle x_1, \ldots, x_n \rangle$, where each x_k corresponds to the number of occurrences of t_k in m. Moreover, each message m of category c_i can be interpreted as the result of picking independently $|m|$ terms from Φ with replacement and probability $P(t_k|c_i)$ for each t_k [21]. Hence, $P(\mathbf{x}|c_i)$ is the multinomial distribution:

$$P(\mathbf{x}|c_i) = P(|m|).|m|!.\prod_{k=1}^{n} \frac{P(t_k|c_i)^{x_k}}{x_k!}.$$

Thus, the criterion for classifying a message as spam becomes:

$$\frac{P(c_s).\prod_{k=1}^{n} P(t_k|c_s)^{x_k}}{\sum_{c_i \in \{c_s, c_l\}} P(c_i).\prod_{k=1}^{n} P(t_k|c_i)^{x_k}} > T,$$

and the probabilities $P(t_k|c_i)$ are estimated as a Laplacian prior:

$$P(t_k|c_i) = \frac{1 + N_{t_k,c_i}}{n + N_{c_i}},$$

where N_{t_k,c_i} is the number of occurrences of term t_k in the training messages of category c_i, and $N_{c_i} = \sum_{k=i}^{n} N_{t_k,c_i}$.

4.3 Multinomial Boolean Naïve Bayes

The multinomial Boolean NB (MN Boolean NB) is similar to the MN TF NB, including the estimation of $P(t_k|c_i)$, but each attribute x_k is Boolean. Note that, these approaches do not take into account the absence of terms ($x_k = 0$) from the messages.

Schneider [25] demonstrates that MN Boolean NB may perform better than MN TF NB. The main reason is that the multinomial NB with term frequency attributes is equivalent to a NB version with the attributes modelled as following Poisson distributions in each category, assuming that the message length is independent of the category. Therefore, the multinomial NB may achieve better performance with Boolean attributes, if the term frequencies attributes do not follow Poisson distributions.

4.4 Multivariate Bernoulli Naïve Bayes

The multivariate Bernoulli NB (MV Bernoulli NB) represents each message m by computing the presence and absence of each term. Therefore, m can be represented as a binary vector $\mathbf{x} = \langle x_1, \ldots, x_n \rangle$, where each x_k shows whether or not t_k will occur in m. Moreover, each message m of category c_i is seen as the result of n Bernoulli trials, where at each trial we decide whether or not t_k will appear in m. The probability of a positive outcome at trial k is $P(t_k|c_i)$. Then, the probabilities $P(\mathbf{x}|c_i)$ are computed by:

$$P(\mathbf{x}|c_i) = \prod_{k=1}^{n} P(t_k|c_i)^{x_k}.(1 - P(t_k|c_i))^{(1-x_k)}.$$

The criterion for classifying a message as spam becomes:

$$\frac{P(c_s).\prod_{k=1}^{n} P(t_k|c_s)^{x_k}.(1 - P(t_k|c_s))^{(1-x_k)}}{\sum_{c_i \in \{c_s,c_l\}} P(c_i).\prod_{k=1}^{n} P(t_k|c_i)^{x_k}.(1 - P(t_k|c_i))^{(1-x_k)}} > T,$$

and probabilities $P(t_k|c_i)$ are estimated as a Laplacian prior:

$$P(t_k|c_i) = \frac{1 + |\mathcal{M}_{t_k,c_i}|}{2 + |\mathcal{M}_{c_i}|},$$

where $|\mathcal{M}_{t_k,c_i}|$ is the number of training messages of category c_i that comprise the term t_k, and $|\mathcal{M}_{c_i}|$ is the total number of training messages of category c_i. For more theoretical explanation, consult Metsis *et al.* [22], or Losada and Azzopardi [18].

4.5 Boolean Naïve Bayes

We denote as Boolean NB the classifier similar to the MV Bernoulli NB with the difference that it does not take into account the absence of terms. Hence, the probabilities $P(\mathbf{x}|c_i)$ are estimated only by:

$$P(\mathbf{x}|c_i) = \prod_{k=1}^{n} P(t_k|c_i),$$

and the criterion for classifying a message as spam becomes:

$$\frac{P(c_s).\prod_{k=1}^{n} P(t_k|c_s)}{\sum_{c_i \in \{c_s,c_l\}} P(c_i).\prod_{k=1}^{n} P(t_k|c_i)} > T,$$

where probabilities $P(t_k|c_i)$ are estimated in the same way as used in the MV Bernoulli NB.

4.6 Multivariate Gauss Naïve Bayes

Multivariate Gauss NB (MV Gauss NB) uses real-valued attributes by assuming that each attribute follows a Gaussian distribution $g(x_k; \mu_{k,c_i}, \sigma_{k,c_i})$ for each category c_i, where the μ_{k,c_i} and σ_{k,c_i} of each distribution are estimated from the training set $\mathcal{M}$.

The probabilities $P(\mathbf{x}|c_i)$ are calculated by

$$P(\mathbf{x}|c_i) = \prod_{k=1}^{n} g(x_k; \mu_{k,c_i}, \sigma_{k,c_i}),$$

and the criterion for classifying a message as spam becomes:

$$\frac{P(c_s).\prod_{k=1}^{n} g(x_k; \mu_{k,c_s}, \sigma_{k,c_s})}{\sum_{c_i \in \{c_s,c_l\}} P(c_i).\prod_{k=1}^{n} g(x_k; \mu_{k,c_i}, \sigma_{k,c_i})} > T.$$

4.7 Flexible Bayes

Flexible Bayes (FB) works similar to MV Gauss NB. However, instead of using a single normal distribution for each attribute X_k per category c_i, FB represents the probabilities $P(\mathbf{x}|c_i)$ as the average of L_{k,c_i} normal distributions with different values for μ_{k,c_i}, but the same one for σ_{k,c_i}:

$$P(x_k|c_i) = \frac{1}{L_{k,c_i}} \sum_{l=1}^{L_{k,c_i}} g(x_k; \mu_{k,c_i,l}, \sigma_{c_i}),$$

where L_{k,c_i} is the amount of different values that the attribute X_k has in the training set $\mathscr{M}$ of category c_i. Each of these values is used as $\mu_{k,c_i,l}$ of a normal distribution of the category c_i. However, all distributions of a category c_i are taken to have the same $\sigma_{c_i} = \frac{1}{\sqrt{|\mathscr{M}_{c_i}|}}$.

The distribution of each category becomes narrower as more training messages of that category are accumulated. By averaging several normal distributions, FB can approximate the true distributions of real-valued attributes more closely than the MV Gauss NB when the assumption that attributes follow normal distribution is violated. For further details, consult John and Langley [14] or Androutsopoulos *et al.* [6].

Table 1 summarizes all NB spam filters presented in this section[3].

5 Support Vector Machines

Support vector machine (SVM) is one of the most successful techniques used in text classification [9, 11]. In this method a data point is viewed as a p-dimensional vector and the approach aims to separate such points with a $(p-1)$-dimensional hyperplane. This is called a linear classifier. There are many hyperplanes that might classify the data. One reasonable choice as the best hyperplane is the one that represents the largest separation, or margin, between the two classes. Therefore, SVM chooses the hyperplane so that the distance from it to the nearest data point on each

[3] The computational complexities are according to Metsis *et al.* [22]. At classification time, the complexity of FB is $O(n.|\mathscr{M}|)$ because it needs to sum the L_k distributions.

Table 1 Naïve Bayes spam filters.

NB Classifier	$P(\mathbf{x}\|c_i)$	Complexity on Training	Classification
Basic NB	$\prod_{k=1}^{n} P(t_k\|c_i)$	$O(n.\|\mathscr{M}\|)$	$O(n)$
MN TF NB	$\prod_{k=1}^{n} P(t_k\|c_i)^{x_k}$	$O(n.\|\mathscr{M}\|)$	$O(n)$
MN Boolean NB	$\prod_{k=1}^{n} P(t_k\|c_i)^{x_k}$	$O(n.\|\mathscr{M}\|)$	$O(n)$
MV Bernoulli NB	$\prod_{k=1}^{n} P(t_k\|c_i)^{x_k}.(1-P(t_k\|c_i))^{(1-x_k)}$	$O(n.\|\mathscr{M}\|)$	$O(n)$
Boolean NB	$\prod_{k=1}^{n} P(t_k\|c_i)$	$O(n.\|\mathscr{M}\|)$	$O(n)$
MV Gauss NB	$\prod_{k=1}^{n} g(x_k;\mu_{k,c_i},\sigma_{k,c_i})$	$O(n.\|\mathscr{M}\|)$	$O(n)$
Flexible Bayes	$\prod_{k=1}^{n} \frac{1}{L_{k,c_i}} \sum_{l=1}^{L_{k,c_i}} g(x_k;\mu_{k,c_i,l},\sigma_{c_i})$	$O(n.\|\mathscr{M}\|)$	$O(n.\|\mathscr{M}\|)$

side is maximized. If such a hyperplane exists, it is known as the maximum-margin hyperplane and the linear classifier it defines is known as a maximum margin classifier (Figure 1) [30].

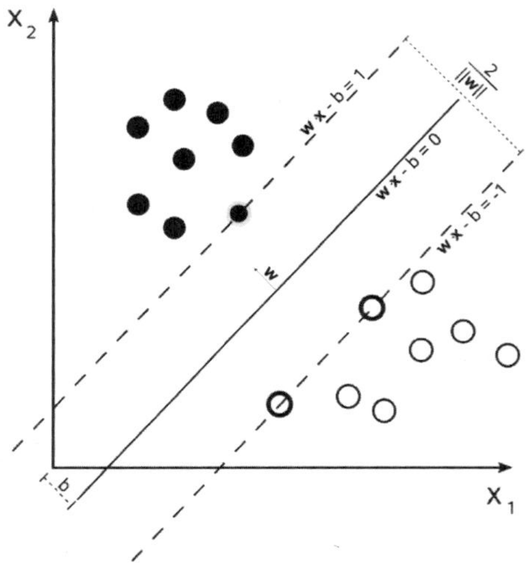

Fig. 1 Maximum-margin hyperplane and margins for a SVM trained with samples from two classes [1].

SVMs belong to a family of generalized linear classifiers. A special property is that they simultaneously minimize the empirical classification error and maximize the geometric margin; hence they are also known as maximum margin classifiers. For further details about the implementation of SVMs in spam filtering domain, consult Cormack [9], Drucker *et al.* [10], Hidalgo [13], Kolcz and Alspector [15], Sculley and Wachman [26], Sculley *et al.* [27] and Liu and Cui [17].

6 Experimental Results

We performed this study on the six well-known, large, real and public Enron datasets[4]. The corpora are composed of legitimate messages extracted from the mailboxes of six former employees of the Enron Corporation. For further details about the dataset statistics and composition, refer to Metsis *et al.* [22].

Tables 2, 3, 4, 5, 6, and 7 present the performance achieved by each classifier for each Enron dataset. Bold values indicate the highest score achieved for the most important metrics. In order to provide a fair evaluation, we consider the most important measures the Matthews correlation coefficient (*MCC*) [1–5] and the weighted accuracy rate ($Acc_w\%$) [6] achieved by each filter. Additionally, we present other well-known measures as spam recall ($Sre\%$), legitimate recall ($Lre\%$), spam precision ($Spr\%$), legitimate precision ($Lpr\%$), and total cost ratio (TCR) [6]. It is important to note that TCR offers an indication of the improvement provided by the filter. A greater TCR indicates better performance, and for $TCR < 1$, it is better to use no filter. On the other hand, the MCC returns a real value between -1 and $+1$. A coefficient equals to $+1$ indicates a perfect prediction; 0, an average random prediction; and -1, an inverse prediction. It can be calculated using the following equation [7, 20]:

$$MCC = \frac{(|\mathscr{TP}|.|\mathscr{TN}|) - (|\mathscr{FP}|.|\mathscr{FN}|)}{\sqrt{(|\mathscr{TP}|+|\mathscr{FP}|).(|\mathscr{TP}|+|\mathscr{FN}|).(|\mathscr{TN}|+|\mathscr{FP}|).(|\mathscr{TN}|+|\mathscr{FN}|)}},$$

where $|\mathscr{TP}|$, $|\mathscr{FP}|$, $|\mathscr{TN}|$ and $|\mathscr{FN}|$ correspond to the amount of true positives, false positives, true negatives and false negatives, respectively.

Table 2 Enron 1 – Results achieved by each filter

Measures	Basic	Bool	MN TF	MN Bool	MV Bern	MV Gauss	Flex Bayes	SVM	MDL
$Sre(\%)$	91.33	96.00	82.00	82.67	72.00	78.67	87.33	83.33	92.00
$Spr(\%)$	85.09	51.61	75.00	62.00	61.71	87.41	86.18	87.41	92.62
$Lre(\%)$	93.48	63.32	88.86	79.35	81.79	95.38	94.29	95.11	97.01
$Lpr(\%)$	96.36	97.49	92.37	91.82	87.76	91.64	94.81	93.33	96.75
$Acc_w(\%)$	92.86	72.78	86.87	80.31	78.96	90.54	92.28	91.70	**95.56**
TCR	4.054	1.064	2.206	1.471	1.376	3.061	3.750	3.488	6.552
MCC	0.831	0.540	0.691	0.578	0.516	0.765	0.813	0.796	**0.892**

Regarding the results achieved by the classifiers, the MDL spam filter outperformed the other classifiers for the majority e-mail collections used in our empirical evaluation. It is important to realize that in some situations the MDL performs

[4] The Enron datasets are available at
`http://www.iit.demokritos.gr/skel/i-config/`

Table 3 Enron 2 – Results achieved by each filter

Measures	Basic	Bool	MN TF	MN Bool	MV Bern	MV Gauss	Flex Bayes	SVM	MDL
$Sre(\%)$	80.00	95.33	75.33	74.00	65.33	62.67	68.67	90.67	91.33
$Spr(\%)$	97.57	81.25	96.58	98.23	81.67	94.95	98.10	90.67	99.28
$Lre(\%)$	99.31	92.45	99.08	99.54	94.97	98.86	99.54	96.80	99.77
$Lpr(\%)$	93.53	98.30	92.13	91.77	88.87	88.52	90.25	96.80	97.10
$Acc_w(\%)$	94.38	93.19	93.02	93.02	87.39	89.61	91.65	95.23	**97.61**
TCR	4.545	3.750	3.659	3.659	2.027	2.459	3.061	5.357	10.714
MCC	0.850	0.836	0.812	0.814	0.652	0.717	0.776	0.875	**0.937**

Table 4 Enron 3 – Results achieved by each filter

Measures	Basic	Bool	MN TF	MN Bool	MV Bern	MV Gauss	Flex Bayes	SVM	MDL
$Sre(\%)$	57.33	99.33	57.33	62.00	100.00	52.67	52.00	91.33	90.00
$Spr(\%)$	100.00	99.33	100.00	100.00	84.75	89.77	96.30	96.48	100.00
$Lre(\%)$	100.00	99.75	100.00	100.00	93.28	97.76	99.25	98.76	100.00
$Lpr(\%)$	86.27	99.75	86.27	87.58	100.00	84.70	84.71	96.83	96.40
$Acc_w(\%)$	88.41	**99.64**	88.41	89.67	95.11	85.51	86.41	96.74	97.28
TCR	2.344	75.000	2.344	2.632	5.556	1.875	2.000	8.333	10.000
MCC	0.703	**0.991**	0.703	0.737	0.889	0.613	0.644	0.917	0.931

Table 5 Enron 4 – Results achieved by each filter

Measures	Basic	Bool	MN TF	MN Bool	MV Bern	MV Gauss	Flex Bayes	SVM	MDL
$Sre(\%)$	94.67	98.00	93.78	96.89	98.22	94.44	94.89	98.89	97.11
$Spr(\%)$	100.00	100.00	100.00	100.00	100.00	100.00	100.00	100.00	100.00
$Lre(\%)$	100.00	100.00	100.00	100.00	100.00	100.00	100.00	100.00	100.00
$Lpr(\%)$	86.21	94.34	84.27	91.46	94.94	85.71	86.71	96.77	92.02
$Acc_w(\%)$	96.00	98.50	95.33	97.67	98.67	95.83	96.17	**99.17**	97.83
TCR	18.750	50.000	16.071	32.143	56.250	18.000	19.565	90.00	34.615
MCC	0.903	0.962	0.889	0.941	0.966	0.900	0.907	**0.978**	0.945

much better than SVM and NB classifiers. For instance, for Enron 1 (Table 2), MDL achieved spam recall rate equal to 92% while SVM attained 83.33%, even thought MDL presented better legitimate recall. It means that for Enron 1 MDL was able to recognize more than 8% of spams than SVM, representing an improvement of 10.40%. In a real situation, this difference would be extremely important. Note that, the same result can be found for Enron 2 (Table 3), Enron 5 (Table 6), and Enron

Table 6 Enron 5 – Results achieved by each filter

Measures	Basic	Bool	MN TF	MN Bool	MV Bern	MV Gauss	Flex Bayes	SVM	MDL
$Sre(\%)$	89.67	87.23	88.86	94.29	98.10	86.68	88.86	89.40	99.73
$Spr(\%)$	98.80	100.00	100.00	100.00	92.56	96.37	98.79	99.70	98.39
$Lre(\%)$	97.33	100.00	100.00	100.00	80.67	92.00	97.33	99.33	96.00
$Lpr(\%)$	79.35	76.14	78.53	87.72	94.53	73.80	78.07	79.26	99.31
$Acc_w(\%)$	91.89	90.93	92.08	95.95	93.05	88.22	91.31	92.28	**98.65**
TCR	8.762	7.830	8.976	17.524	10.222	6.033	8.178	9.200	52.571
MCC	0.825	0.815	0.835	0.909	0.828	0.743	0.814	0.837	**0.967**

Table 7 Enron 6 – Results achieved by each filter

Measures	Basic	Bool	MN TF	MN Bool	MV Bern	MV Gauss	Flex Bayes	SVM	MDL
$Sre(\%)$	86.00	66.89	76.67	92.89	96.22	92.00	89.78	89.78	98.67
$Spr(\%)$	98.98	99.67	99.42	97.21	92.32	94.95	98.30	95.28	95.48
$Lre(\%)$	97.33	99.33	98.67	92.00	76.00	85.33	95.33	86.67	86.00
$Lpr(\%)$	69.86	50.00	58.50	81.18	87.02	78.05	75.66	73.86	95.56
$Acc_w(\%)$	88.33	75.00	82.17	92.67	91.17	90.33	91.17	90.05	**95.50**
TCR	6.716	3.000	4.206	10.227	8.491	7.759	8.491	6.818	16.667
MCC	0.757	0.574	0.661	0.816	0.757	0.751	0.793	0.727	**0.878**

6 (Table 7). Both methods, MDL and SVM, achieved similar performance with no significant statistical difference just for Enron 3 (Table 4) and Enron 4 (Table 5).

The results indicate that the data compression model is more efficient to distinguish messages as spams or legitimates. It attained an accuracy rate higher than 95% and high precision × recall rates for all datasets indicating that the MDL classifier makes few mistakes. We also verify that the MDL classifier achieved *MCC* score higher than 0.87 for all tested corpus. It indicates that the proposed filter almost accomplished a perfect prediction ($MCC = 1.000$) and it is much better than not using a filter ($MCC = 0.000$).

Among the evaluated NB classifiers, the results indicate that all of them achieved similar performance with no significant statistical difference. However, they had achieved lower results than MDL and linear SVM, which attained accuracy rate higher than 90% for all Enron datasets.

Moreover, according to the results found by Schneider [25], in our experiments the NB filters that use real and integer attributes did not achieved better results than Boolean ones. However, Metsis *et al.* [22] showed that flexible Bayes are less sensitive to the threshold T. It indicates that it is able to attain a high spam recall even though a high legitimate recall is required.

7 Conclusions

In this paper, we have presented a new spam filtering approach based on the Minimum Description Length principle. We have also compared its performance with the linear Support Vector Machine and seven different models of Naïve Bayes classifiers, something the spam literature does not always acknowledge.

We have conducted an empirical experiment using six well-known, large, and public databases and the reported results indicate that the proposed classifier outperforms currently established spam filters. It is important to emphasize that MDL spam filter acquired the best average performance for all analyzed databases presenting an accuracy rate higher than 95% for all e-mail datasets.

Actually, we are conducting more experiments using larger datasets as TREC05, TREC06, and TREC07 corpora [9] in order to reinforce the validation. We also intend to compare the approaches with other commercial and open-source spam filters, such as Bogofilter, SpamAssassin, OSBF-Lua, among others.

Future researches should take into consideration that spam filtering is a coevolutionary problem, because while the filter tries to evolve its prediction capacity, the spammers try to evolve their spam messages in order to overreach the classifiers. Hence, an efficient approach should have an effective way to adjust its rules in order to detect the changes of spam features. In this way, collaborative filters [16] could be used to assist the classifier by accelerating the adaptation of the rules and increasing the classifiers' performance. Moreover, spammers generally insert a large amount of noise in spam messages in order to make the probability estimation more difficult. Thus, the filters should have a flexible way to compare the terms in the classifying task. Approaches based on fuzzy logic [31] could be employed to make the comparison and selection of terms more flexible.

Acknowledgements. The authors would like to thank J. Almeida for his very constructive suggestions and the Brazilian Coordination for the Improvement of Higher Level Personnel (Capes) for financial support.

References

[1] Almeida, T., Yamakami, A.: Content-Based Spam Filtering. In: Proceedings of the 23rd IEEE International Joint Conference on Neural Networks, Barcelona, Spain, pp. 1–7 (2010)

[2] Almeida, T., Yamakami, A., Almeida, J.: Evaluation of Approaches for Dimensionality Reduction Applied with Naive Bayes Anti-Spam Filters. In: Proceedings of the 8th IEEE International Conference on Machine Learning and Applications, Miami, FL, USA, pp. 517–522 (2009)

[3] Almeida, T., Yamakami, A., Almeida, J.: Filtering Spams using the Minimum Description Length Principle. In: Proceedings of the 25th ACM Symposium On Applied Computing, Sierre, Switzerland, pp. 1856–1860 (2010a)

[4] Almeida, T., Yamakami, A., Almeida, J.: Probabilistic Anti-Spam Filtering with Dimensionality Reduction. In: Proceedings of the 25th ACM Symposium On Applied Computing, Sierre, Switzerland, pp. 1804–1808 (2010b)

[5] Almeida, T., Almeida, J., Yamakami, A.: How the Dimensionality Reduction Affects the Accuracy of Naive-Bayes Classifiers. Journal of Internet Services and Applications 1(3), 183–200 (2011)

[6] Androutsopoulos, I., Paliouras, G., Michelakis, E.: Learning to Filter Unsolicited Commercial E-Mail. Technical Report 2004/2, National Centre for Scientific Research "Demokritos", Athens, Greece (2004)

[7] Baldi, P., Brunak, S., Chauvin, Y., Andersen, C., Nielsen, H.: Assessing the Accuracy of Prediction Algorithms for Classification: An Overview. Bioinformatics 16(5), 412–424 (2000)

[8] Barron, A., Rissanen, J., Yu, B.: The Minimum Description Length Principle in Coding and Modeling. IEEE Transactions on Information Theory 44(6), 2743–2760 (1998)

[9] Cormack, G.: Email Spam Filtering: A Systematic Review. Foundations and Trends in Information Retrieval 1(4), 335–455 (2008)

[10] Drucker, H., Wu, D., Vapnik, V.: Support Vector Machines for Spam Categorization. IEEE Transactions on Neural Networks 10(5), 1048–1054 (1999)

[11] Forman, G., Scholz, M., Rajaram, S.: Feature Shaping for Linear SVM Classifiers. In: Proceedings of the 15th ACM SIGKDD International Conference on Knowledge Discovery and Data Mining, Paris, France, pp. 299–308 (2009)

[12] Grünwald, P.: A Tutorial Introduction to the Minimum Description Length Principle. In: Grünwald, P., Myung, I., Pitt, M. (eds.) Advances in Minimum Description Length: Theory and Applications, pp. 3–81. MIT Press (2005)

[13] Hidalgo, J.: Evaluating Cost-Sensitive Unsolicited Bulk Email Categorization. In: Proceedings of the 17th ACM Symposium on Applied Computing, Madrid, Spain, pp. 615–620 (2002)

[14] John, G., Langley, P.: Estimating Continuous Distributions in Bayesian Classifiers. In: Proceedings of the 11st International Conference on Uncertainty in Artificial Intelligence, Montreal, Canada, pp. 338–345 (1995)

[15] Kolcz, A., Alspector, J.: SVM-based Filtering of E-mail Spam with Content-Specific Misclassification Costs. In: Proceedings of the 1st International Conference on Data Mining, San Jose, CA, USA, pp. 1–14 (2001)

[16] Lemire, D.: Scale and Translation Invariant Collaborative Filtering Systems. Information Retrieval 8(1), 129–150 (2005)

[17] Liu, S., Cui, K.: Applications of Support Vector Machine Based on Boolean Kernel to Spam Filtering. Modern Applied Science 3(10), 27–31 (2009)

[18] Losada, D., Azzopardi, L.: Assessing Multivariate Bernoulli Models for Information Retrieval. ACM Transactions on Information Systems 26(3), 1–46 (2008)

[19] Marsono, M., El-Kharashi, N., Gebali, F.: Targeting Spam Control on Middleboxes: Spam Detection Based on Layer-3 E-mail Content Classification. Computer Networks 53(6), 835–848 (2009)

[20] Matthews, B.: Comparison of the Predicted and Observed Secondary Structure of T4 Phage Lysozyme. Biochimica et Biophysica Acta 405(2), 442–451 (1975)

[21] McCallum, A., Nigam, K.: A Comparison of Event Models for Naive Bayes Text Classication. In: Proceedings of the 15th AAAI Workshop on Learning for Text Categorization, Menlo Park, CA, USA, pp. 41–48 (1998)

[22] Metsis, V., Androutsopoulos, I., Paliouras, G.: Spam Filtering with Naive Bayes - Which Naive Bayes? In: Proceedings of the 3rd International Conference on Email and Anti-Spam, Mountain View, CA, USA, pp. 1–5 (2006)
[23] Rissanen, J.: Modeling by Shortest Data Description. Automatica 14, 465–471 (1978)
[24] Sahami, M., Dumais, S., Hecherman, D., Horvitz, E.: A Bayesian Approach to Filtering Junk E-mail. In: Proceedings of the 15th National Conference on Artificial Intelligence, Madison, WI, USA, pp. 55–62 (1998)
[25] Schneider, K.-M.: On word frequency information and negative evidence in naive bayes text classification. In: Vicedo, J.L., Martínez-Barco, P., Muńoz, R., Saiz Noeda, M. (eds.) EsTAL 2004. LNCS (LNAI), vol. 3230, pp. 474–485. Springer, Heidelberg (2004)
[26] Sculley, D., Wachman, G.: Relaxed Online SVMs for Spam Filtering. In: Proceedings of the 30th International ACM SIGIR Conference on Research and Development in Information Retrieval, Amsterdam, The Netherlands, pp. 415–422 (2007)
[27] Sculley, D., Wachman, G., Brodley, C.: Spam Filtering using Inexact String Matching in Explicit Feature Space with On-Line Linear Classifiers. In: Proceedings of the 15th Text REtrieval Conference, Gaithersburg, MD, USA, pp. 1–10 (2006)
[28] Siefkes, C., Assis, F., Chhabra, S., Yerazunis, W.S.: Combining winnow and orthogonal sparse bigrams for incremental spam filtering. In: Boulicaut, J.-F., Esposito, F., Giannotti, F., Pedreschi, D. (eds.) PKDD 2004. LNCS (LNAI), vol. 3202, pp. 410–421. Springer, Heidelberg (2004)
[29] Song, Y., Kolcz, A., Gilez, C.: Better Naive Bayes Classification for High-precision Spam Detection. Software – Practice and Experience 39(11), 1003–1024 (2009)
[30] Vapnik, V.N.: The Nature of Statistical Learning Theory. Springer, New York (1995)
[31] Zadeh, L.: Fuzzy sets. Information and Control 8(3), 338–353 (1965)
[32] Zhang, L., Zhu, J., Yao, T.: An Evaluation of Statistical Spam Filtering Techniques. ACM Transactions on Asian Language Information Processing 3(4), 243–269 (2004)

Computational Intelligent Techniques and Similarity Measures for Malware Classification

M.K. Shankarpani, K. Kancherla, R. Movva, and S. Mukkamala

Abstract. One of the major problems concerning information security is malicious code. To evade detection, malware (unwanted malicious piece of code) is packed, encrypted, and obfuscated to produce variants that continue to plague properly defended and patched systems and networks with zero day exploits. Zero day exploits are used by the attackers to compromise victims computer before the developer of the target software knows about the vulnerability.

In this chapter we present a method of functionally classifying malicious code that might lead to automated attacks and intrusions using computational intelligent techniques and similarity measures. We study the performance of kernel methods in the context of robustness and generalization capabilities of malware classification.

Results from our recent experiments indicate that similarity measures can be utilized to determine the likelihood that a piece of code or binary under inspection contains a particular malware. Malware variants of a particular malware family show very high similarity scores (over 85%). Interestingly Trojans and hacking tools have high similarity scores with other Trojans and hacking tools.

Our results also show that malware analysis based on the API calling sequence and API frequency that reflects the behavior of a particular piece of code gives good accuracy to classify malware. We also show that classification accuracy varies with the kernel type and the parameter values; thus, with appropriately chosen parameter values, malware can be detected by support vector machines (SVM) with higher accuracy and lower rates of false alarms.

M.K. Shankarpani · K. Kancherla · R. Movva · S. Mukkamala
Institute for Complex Additive Systems Analysis
Computational Analysis and Network Enterprise Solutions, CAaNES
New Mexico Tech
Socorro, U.S.A.
e-mail: {madhuk,kancherla,srinivas}@cs.nmt.edu,
ram@cybersecurityworks.com

D.A. Elizondo et al. (Eds.): Computational Intelligence for Privacy and Security, SCI 394, pp. 215–236.
springerlink.com

1 Introduction

The Internet is often characterized as a public, interconnected, distributed, open, and dynamic socio-economic playground for cybercriminals. The phenomenal growth of computing devices, connectivity speed, and number of applications running on networked systems pose engendering risk to the Internet. Today, majority of the computers across the globe are interconnected using a multitude of platform independent programs and are becoming increasingly popular. Proliferation of malicious code and malware to these interconnected systems is posing a serious threat to individual users, and enterprises alike.

In modern computer systems, malware is evolving as a major threat that is rapidly increasing. To identify these malicious programs, the currently used signatures and anomaly-based techniques are turning out to be obsolete. To avoid detection, malware authors are developing stealth techniques that are built into malware. To conquer the drawbacks of signature based algorithms in detecting these new breeds of malware, the focus of malware research has shifted towards finding more generalized and scalable features to identify malware, rather than relying only on signatures.

1.1 Malware Attack Toolkits

Attack toolkits are groups of malicious programs that are used to facilitate widespread attacks on computer networks [1]. The primary objective of these prewritten toolkits is to exploit vulnerabilities that typically steal sensitive data or to convert the compromised computers into botnets for additional attacks. These kits are sold in underground economies, where cybercriminals trade stolen information and services. These attack kits play a significant role in the evolution of cybercrime as a self-sustaining underground economy with the motive, of generating vast profits worth millions of dollars. Over the last decade, the drastic increase in number of toolkits developed and released in the wild signifies that the malware developers are making profit from these attack kits.

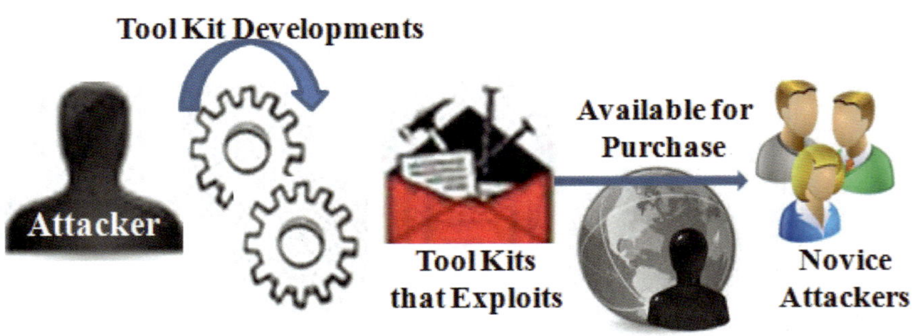

Fig. 1 Creation and sales of attack tool kits.

Figure 1 illustrates the methodology adopted by an attacker in developing attack tool kits which include exploits for various vulnerabilities. The driving force for evolution of attack kits is the ability to facilitate widespread attacks for profit. Even the end users who purchase these toolkits need not necessarily have an in-depth knowledge of attack concepts; thus a novice attacker can launch a sophisticated attack without knowing the vulnerability or how to exploit it. These kits are more marketable and easier to deploy because of the well-developed and easy to use graphical user interfaces.

1.2 Attack Methods

Numerous vulnerabilities on the victim's computer lead to the growth of attack kits. These are a serious threat to organizations and end users since they aid and abet novice cybercriminals in exploiting these vulnerabilities and facilitate in installation of malicious code on compromised computers. Increasingly, many kits are available with a wide range of exploits and a large array of attack vectors. These kits are hosted on malicious websites for launching attacks. Web browsers and browser plug-ins are the prime targets of these exploits. Attacking client-side vulnerabilities will minimize the attack footprint, and also increase the likelihood to penetrate through the perimeter defense.

In these kinds of attacks, the attacker creates malicious websites that appear legitimate to visitors, but contain hidden exploits that targets user computers. Social engineered attacks are also among several attack methods used by attackers to compromise systems. In this method, the user is enticed into clicking on malicious links or downloading files like PDFs, codec support for streaming videos, etc., which are usually programs that exploit the vulnerabilities on the user's computer.

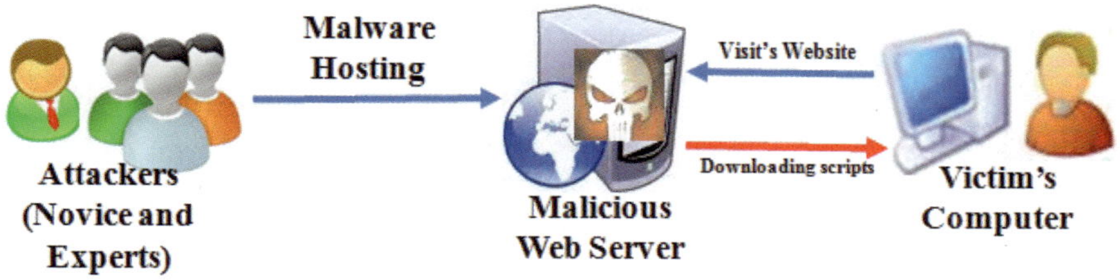

Fig. 2 Usage of attack tool kits to exploit victim's computer

In Figure 2 the attacker can target the victim's computer in different ways including but not restricted to phishing, social engineering, email attachments, etc. Among all these the most common method is to host the tool kits on malicious websites, as it is the easiest way to subvert perimeter defense controls like firewalls, Intrusion Detection and Prevention Systems (IDPSs), etc. When the user visits or gets redirected to these websites; malware downloads malicious scripts and installs them on the user's computer without the user's consent; thus taking complete control of the victim's computer by exploiting its vulnerabilities.

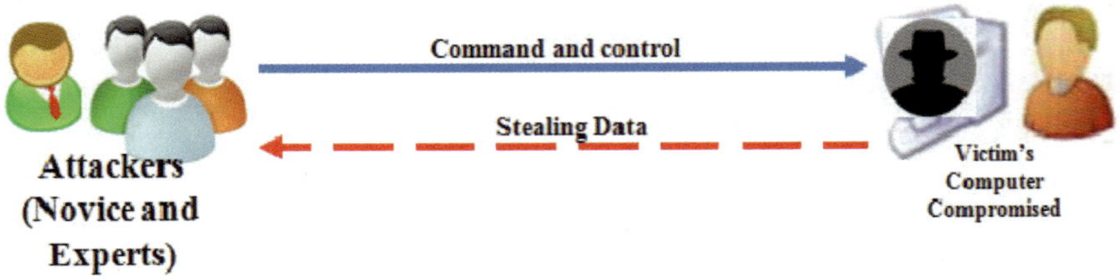

Fig. 3 Attackers have access to victim's computer

When the user's computer is compromised as shown in Figure 3, the victim's activities are monitored and sensitive data is transferred to the attacker without user's knowledge. Sensitive data might consist of personal identifiable information like bank accounts, user credentials, SSN, date of birth, and location etc. Depending on the attacker's intention, command and control (C&C) is used either to steal user data from the victim's computer or the victim's machine will be made part of botnets to launch attacks on other computers.

1.3 Anti-antivirus Methodologies

Malware authors have been working on anti-antivirus techniques and are creating increasingly complex malware. We explain a few methodologies developed by malware writers intended to bypass the antivirus systems.

- **Code Obfuscation:** In this technique, the malware code includes unnecessary jumps, no-op instructions, replacing unused registers, simple and dummy loops etc, so that the code looks tangled and the signature based detection fails.
- **Encryption:** Encryption is one of the easiest ways to hide malware functionality. The encrypted malware contains an encrypted malicious body and an unencrypted decryption engine. For each variant the encryption is done with unique key which easily defeats the detection systems using unique virus signatures.
- **Polymorphism:** Antivirus systems detect encrypted viruses by searching for signatures of the decryption engine which is not encrypted. As mutating techniques emerged, polymorphic malware came into existence. In polymorphic malware, the payload is packed, which gets unpacked on the fly. Once the unpacking is complete, the payload instructions get executed accordingly.
- **Metamorphism:** Metamorphic malware do not contain any constant parts, it's a self-mutating malware which changes itself during the infection process and the infected file does not resemble the parent, thus creation of signatures is extremely difficult.

Though heuristic based antivirus systems can be used to detect the polymorphic malware, the learning curve for those kinds of systems is very steep. Mainly it is difficult for heuristic based antivirus systems to predict the normal behavior of the system, due to constant change of the existing computer file system either due to installation of vendor patches or end user activity. In addition, the main drawback of this system is its high rates of false positives.

In this chapter, we present two approaches for malware analysis and its detection namely, support vector machines and similarity analysis approach. We also present the methodology of unpacking a packed executable by using the memory violation exception raised by hardware. Our approach rests on the analysis of the frequency of occurrence of each Windows Application Programming Interface (API).

This chapter shows how to classify and detect a malware based on term frequency of Windows API rather than its sequence of occurrence. In this methodology, data samples of malware that are new and its variants are considered. To determine the classifier accuracy a few benign executables have been included in the data sample.

The organization of this chapter is as follows: Section 2 represents current issues and challenges faced by the existing malware detection systems, with an example of a malware anatomy explaining the vulnerabilities that are exploited by malware. The structure of a portable executable is given in Section 3, followed by our methodology of analyzing a portable executable along with its experimental results in Section 4. Section 5 concludes that our method is very effective against polymorphic or metamorphic type of variants of an executable.

2 Current Issues with Malware Detection Systems

In current scenario malware cannot be considered as sequential programs, but rather as a system which interacts with its environment. Malicious executables are programs which can infiltrate or damage a computer system without the owner's consent and pose a serious threat to the security community. It is highly complex and difficult to detect polymorphic viruses that adopt obfuscation techniques. What is making the situation worse is the relative ease of producing polymorphic (or rapid variants of the original malware) and metamorphic malicious code. Detection of metamorphic malware is even more complex and difficult when compared to its native versions.

2.1 Limitations of Current Malware Detection Systems

New variants may or may not be created by the same author. If the original author releases his or her malicious source code, then other authors can join in creating a plethora of its variants that are undetectable by the current anti-virus software. With increasing use of obfuscation and other complex techniques to release new variants; security companies are forced to constantly update their signature databases by creating a new signature each time a variant or a new malicious code is released.

The process of reverse engineering multilayered packed executables is a daunting task. To make the situation even worse, malicious code once confined to wired networks, has now found a new breeding ground in mobile devices. This includes Automatic Identification and Data Collection (AIDC) technologies like Radio Frequency Identification Devices (RFID) that use wireless networks to communicate over the Internet.

In order to stay ahead of the curve and be proactive in an asymmetric race against malicious code writers, developers of security technologies have to rely on automatic and behavioral analysis tools [2, 3]. The hypothesis is that all versions of malicious code or similar malware share a common core signature that is a combination of several features of the code. After a particular malicious code or malware has been identified for the first time, it can be analyzed to extract the signature, which provides a basis for detecting its variants and mutants in the future.

Present malware detection software uses signature-based and heuristic-based algorithms. Signature-based algorithms create a unique short string of bytes for a given virus; and later use them to detect similar virus patterns in files, present on the disk or on the memory with low error rates [4]. Generally speaking, detection strategies based on string signatures use a database of regular expressions and a string matching engine to scan files and detect infected ones. Each regular expression of the database is designed to identify a known malicious program.

There are at least three impediments associated with this approach. First, the identification of a malware signature requires a human expert investing significantly large amounts of time to forge a reliable signature when compared to the time taken for a malware attack. Secondly, the string signature approach can be easily bypassed by obfuscation techniques. Thirdly, as the quantity of malware increases, the ratio of false positives becomes a crucial factor in its detection.

Signature-based detection is slow and might be ineffective against polymorphic and unknown malware. On the other hand, heuristic-based algorithms detect unknown malware, but they are usually inefficient and inaccurate. The inability of these traditional approaches to detect new forms of malicious executables has shifted the focus of virus research to more generalized and scalable techniques that can identify malicious behavior as a process instead of being detected by a unique signature.

2.2 Related Work

Though traditional signature-based malware detection methods exist, there is large scope for improvement [5, 6] and in order to detect new malware, a few data mining and machine learning techniques are approached. Schultz et al. [7] applied a Naïve Bayes method to detect new unknown malware. Kolter et al. [8] examined the performance of various classifiers such as Naïve Bayes, support vector machine (SVM) and plotting Receiver Operating Characteristic (ROC) curves using decision tree methods. Yanfang Ye et al. [9] applied Objective-Oriented Association (OOA) mining based classification [10, 11] on Windows API execution sequences called by Portable Executable files.

A few of these methods entirely rely on the occurrence of API execution sequence. There are also methods where websites are crawled to inspect if they host any kind of malicious executables [12]. This study is generally for web server security, advertising and third-party widgets. Their basic approach shows how malware executables are often distributed across a large number of URLs and domains. Analyzing and detecting these obfuscated malicious executables is by itself a vast field.

Our work is based on the large collection of real world malicious executables collected at Offensive Computing [13]. As of today, Offensive Computing has one of the largest malware databases which include various kinds of executables including but not limited to spyware, adware, virus, worms, trojans, etc. Among thousands of malware in the computing world, the number of unique executables is likely to be much lower, since many binaries differ only in their binary packing and not in their functionality [14]. In this chapter we show how these obfuscated malicious executables can be detected effectively.

2.3 Illustration of Malware Anatomy

Malware are programs that do not just infect files, but can also exploit existing vulnerabilities on the system. Changes to OS architecture, vulnerability patches or anti-virus updates do not stop the propagation of malware. By taking one of the most prolific computer threats in present times, Zeus and its variants as an example, we discuss how a malware exploits the vulnerabilities on a user computer for its propagation. Because of the vulnerabilities and delay in patching them, malware can easily compromise computers and carry out malicious and unauthorized activities. This section discusses one of the most prolific computer threat in present times Zeus and its variants.

Zeus is considered as one of the most nefarious financial and banking Trojans that primarily targets business and financial institutions to perform unauthorized Automated Clearing House (ACH) and wire transfer transactions for check and payment processing. Zeus has been causing loss of billions of dollars and is facilitating identity theft of innocent users for financial gains.

Zeus crimeware (Crimeware is a group of malware specially designed by the attackers to automate cybercrime) does one thing very well that every security researcher envy's – obfuscation. The Zeus kit conceals the exploit code every time a binary is created. It has an inbuilt binary generator that generates a new binary file on every use that is radically different from others; which evades detection from antivirus or security technologies that rely on signature based detection. **The effectiveness of an up to date antivirus against** Zeus **is thus not 100%, not 90%, not even 50% – it's just 23% which is alarming.**

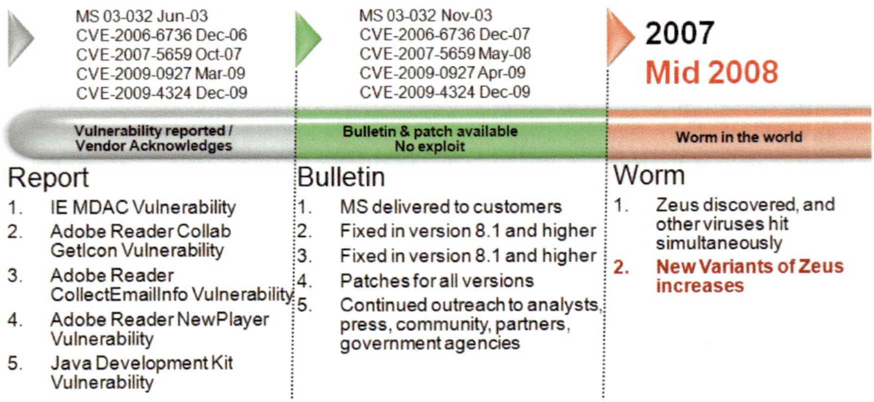

Fig. 4 Anatomy of Zeus and its variants

Figure 4 shows the vulnerabilities that Zeus and a few other malware exploited [15]. The antivirus signatures were very effective and controlled the spread of Zeus and its variants until around 2008, during which the number of variants created was limited. From mid-2008, cybercriminals started to develop Zeus builder, which has the capability to create as many variants as required by the attackers with unique configuration files for each of its variants. Unfortunately, the antivirus companies could not keep up with the pace of creating signatures for detecting those variants. A record number of over 5,079 variants of Zeus were reached by May 2009 [16]. This shows that signature based malware detection techniques are inefficient as the number of variants or new malware created is on the rise. Through this chapter, we put forth a new malware detection technique, by categorizing malware based on their API function calls.

3 Anatomy of Portable Executable

Microsoft introduced the Portable Executable file format, most commonly known as the PE format. The main intention was to develop a common file format to support all windows flavors and on all supported CPUs. The PE32+ format for 64-bit windows has a few modifications compared to the 32-bit PE. No new fields are added and most of the changes are simply widening the fields from 32-bits to 64-bits [17]. The DLL file format is exactly the same as PE with a difference in a single bit which indicates whether the file is an executable or a DLL. This section gives an overview of the PE file format, and its importance in detection of malware.

3.1 Overview of the PE File Format

All code, data and resources of an executable are represented by the module in the memory. Other parts may be read, but not mapped onto memory, such as debug information placed at the end of the file. The PE header provides information on how much memory should be reserved by the computer to map the executable onto memory. In PE, the code section contains the code and the data section contains different types of data, including API import and export tables, resources, and relocations. Each of these sections has its own memory attributes.

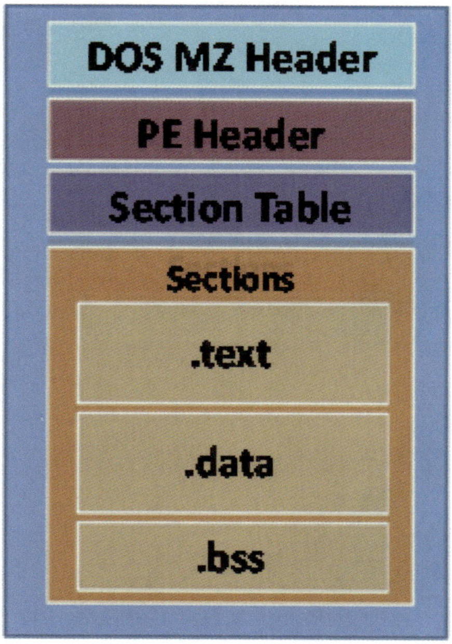

Fig. 5 Microsoft Portable Executable (PE) file format

Figure 5 shows Win32 specifications, which is developed by Microsoft for its PE file format. The most important characteristics of PE are that its data structure remains the same on the disk and when it is loaded onto the memory. Loading an executable onto memory is the process of mapping certain range of a PE file into the address space; however the offset of the file loaded onto memory may differ from the one on the disk.

Windows loader loads the executable onto memory. The executable on the memory is called a module and the starting address location where the file mapping begins is referred as HMODULE, which is very useful in finding other data structures in memory. Let us take a look at the PE format. To show, as an example, a benign Visual C++ program is reverse engineered and studied.

3.1.1 DOS Stub

At the start of the PE we find the DOS stub, which is well known from the 16-bit windows executable. In the PE, the DOS stub contains the most common output error message such as "This program cannot be run in DOS Mode". It also consists of header information like 2 bytes of magic number "MZ", the initials of Mark Zbikowski, which in hexadecimal representation is 0x5A4D.

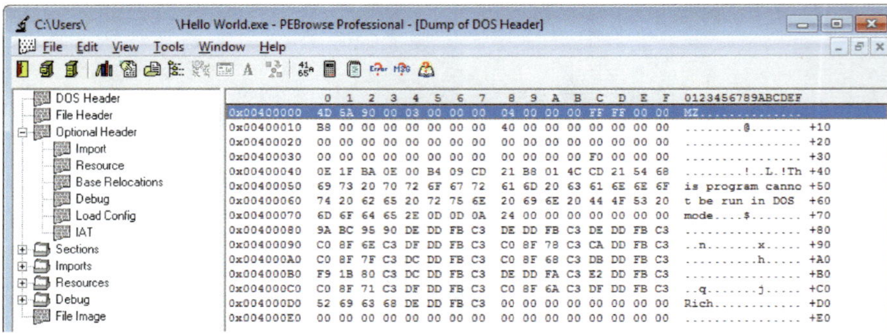

Fig. 6 View of the DOS Header of a normal executable in Hex editor

Figure 6 shows an example of the DOS header information of a normal file. It clearly shows that the first two bytes start with "MZ" and the DOS error message.

3.1.2 The PE Header

The PE Header contains the following fields:

- "Machine type" on which the executable is intended to run, which may contain values such as 0x14C for Intel 80xxx processor, 0x184 for DEC Alpha AXP processor, etc.
- "Number of Sections" in the PE file.
- "Time Date Stamp" when the PE file was created.
- "Characteristics" field contains a 16-bit value that represents whether the file line number is stripped, or whether the file is an executable and other details.

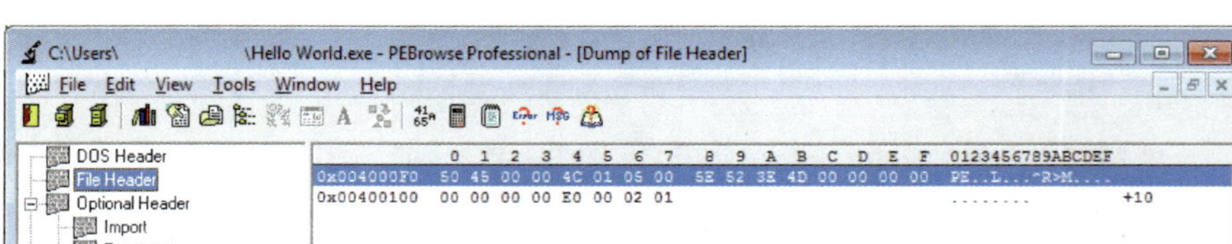

Fig. 7 View of the PE Header of a normal executable file in Hex editor

Figure 7 shows the first 4 bytes reading 0x00004550, and its corresponding ASCII value "PE00", which represents it as a PE file. The next 2 bytes read 0x014C, which represents an i386 machine, the machine type on which it is intended to run.

3.1.3 Section Table

The section table consists of an array of two parts:

- Section Description
- Raw section data

Each section in the array contains the name of the section, its relative virtual address to be loaded onto memory, the size of the raw data in that section and a pointer to the raw data. With reference to Figure 5:

- The ".text" section contains the executable code, which may have other names as well such as ".code", ".AUTO" and others.
- The ".data" section contains all initialized static variables with appropriate memory read/write bits that are set to 1. Usually this section size is computed using the sum of the base of the data and the size of the initialized data. Typical names for this section are ".data", ".idata", "DATA" and others.
- The ".bss" section contains all uninitialized static variables. This section is similar to the ".data" section, but will have a file offset of zero indicating that its contents are not stored in the file.

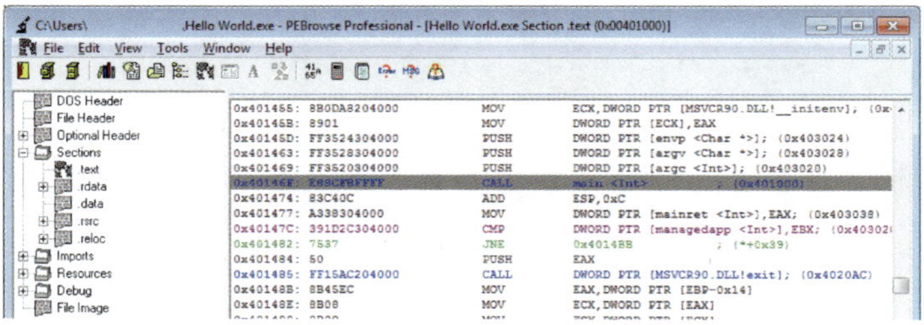

Fig. 8 Executable sections and assembly instructions of a PE

In Figure 8, we can see that the executable contains five sections with names .text, .rdata, .data, .rsrc, .reloc, and each of these sections are categorized based on the memory permissions needed while execution. This approach reduces the time taken for execution, as the permission bits for each section are set prior to execution.

A sample of code section, .text and RVA of the instructions are shown in the above figure. The highlighted instruction on address 0x40146F is the entry point of the executable which is under examination. The entry point refers to the starting of the program which gets executed first in that executable file.

3.1.4 Importing Functions

While using the code of another DLL, we are importing the functions from those DLL(s). When PE files load, the Windows Loader locates and imports all the required functions and data, and makes sure that these addresses are available to the PE file.

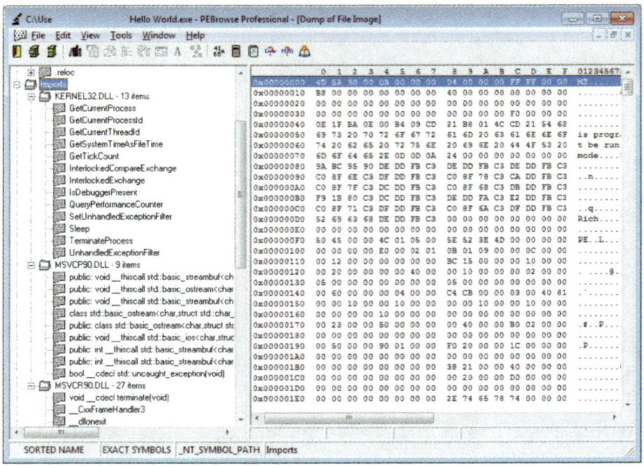

Fig. 9 DLL and its functions imported for execution of a PE file

Figure 9 shows the DLLs and its imported functions for execution of a PE file. These import and export function addresses are maintained in the Import Address Table (IAT). The figure shows functions from KERNEL32.DLL, MSVCP90.DLL and MSVCR90.DLL as loaded by the Windows loader when the PE file was loaded onto memory.

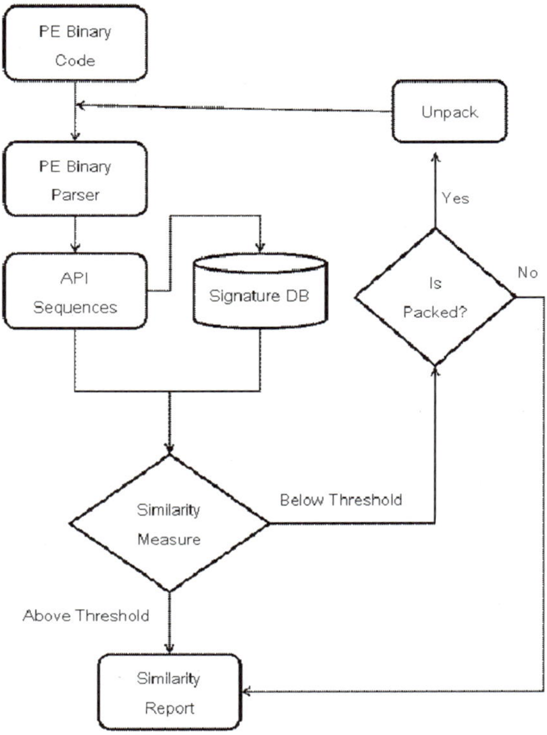

Fig. 10 Flow diagram of similarity analysis

4 Analysis Methodology

In this section we present our methodology of analyzing a suspicious executable. In our approach we unpack the executable automatically and subsequently generate a similarity report. In the later part of this section we show how machine learning techniques can be used to effectively classify and detect a malware.

A pictorial diagram of our methodology is shown in Figure 10. First, the PE code is (optionally) decompressed and passed through a PE file parser, producing the intermediate representation which consists of a Windows API calling sequence. It then maps each Windows API to a global 32-bit integer number. The 16 most significant bits of the integer represent a particular Win32 module (dynamically-linked library), and the last 16 bits represent a particular API in this module. Each API of the calling sequence is represented by its respective global id number that represents the static calling sequence.

This sequence is compared to a known malware sequence or signature (from the signature database) and is passed through the similarity measure module to generate the similarity report. If the similarity value is above a certain threshold (over 70%) then it is classified as malware, if the similarity score is less than (70%) then it is classified as a normal program. The detection decision is made based on the similarity report.

In our methodology, the suspicious PE files are reverse engineered to extract the API call sequence, which are later used for our detection mechanism. If a PE file is packed, using a generic PE parsing method, it is not feasible to extract the complete set of the API call sequence. This may lead to a misclassification of that executable. Hence, we have implemented a generic unpacking methodology to extract the hidden API call sequence. The remaining of this section describes a generic unpacking method used in our experiments.

During the PE parsing process the following errors might indicate that an executable is packed: The Import Address Table (IAT) is corrupted or wrong, has entry point outside the specified code, mismatch between the size of PE file and the size when it is loaded onto the memory, incorrect ordinal names, etc.

Our method of unpacking is based on tracking execution after written memory traces. All memory writes of an executable are traced and if the program counter tries to execute from the written memory address, indicating that some kind of a procedure similar to unpacking, self-modification has occurred.

When a change is likely to occur on virtual mapping or on memory protection, the hardware will raise an exception and pass it to the operating system to handle the exception; we trap this exception to examine the process and reconstruct the API sequence. Finally the API sequences are extracted from the executable and are used for similarity measures.

During the unpacking process, the unpacked code is written to memory and the memory page is marked with write permission but restricted from executable permission. When the programs access the same page to execute the instructions, a memory access violation exception is reported to the operating system. At this juncture, we intercept and handle the exception. Each time the memory protection error is raised by the CPU during the execution of the suspicious program, our algorithm is invoked. This methodology is implemented on a page level granularity rather than instruction level, thus improving the runtime of the algorithm. When our algorithm is invoked, it reanalyzes the suspicious program on the memory and extracts the API calls from the import tables. This API sequence is then used for detection or for creating signature of the suspicious file.

If a suspicious program is packed, we copy all the page permissions before executing the program. When execute-on-write occurs, a page fault exception is raised and suitable action is taken. If the raised page fault exception is other than the one of our interest we pass it on to OS to handle it safely. Later the API call sequence extracted from this process is used for evaluation of similarity measures. If the suspicious file is a malware, its signature is created and updated into our database.

In similarity measurement, the API sequence extracted from the executable is compared with the API sequence of the signature database. The following section explains the methods used to calculate the similarity measure between the two set of APIs and present their results.

4.1 Similarity Analysis Results

A signature is an API sequence of a known virus that has been previously identified. Let's denote it as *Vs* (vector of signature). The API sequence of a suspicious PE binary file is denoted as *Vu* (vector of unknown). To identify whether the new executable with signature *Vu* is an obfuscated version of the virus represented by *Vs*, we measure the similarity between *Vs* and *Vu*.

Our method of detection consists of two steps:

- Aligns sequences of API calls
- Calculates similarity measures using: Cosine, Extended Jaccard, and Pearson

Table 1 Optimal alignment algorithm

	W	A	N	D	E	R
W	**X**					
A		**X**				
D				**X**		
E					**X**	
R						**X**
S						

The optimal alignment algorithm as shown in Table 1 can be conceptualized by considering a matrix with the first sequence placed horizontally at the top and the second sequence placed vertically on the side. Each position in the matrix corresponds to a position in the first and second sequence. Any alignment of the two sequences corresponds to a path through the grid.

Using paths in the grid to represent alignments provides an effective method for computing the best alignment. The score of the best path up to that position is placed in each cell. Beginning at the top left cell, the scores are calculated as the sum of the scores of the element pairs determined by the score of the row and column heading (0 for mismatches and 1 for matches) and the highest score in the grid above and to the left of the cell.

Let us assume "WANDER" and "WADERS" are two API sequences to compute similarity between them. Table 1 shows the path in the grid for best alignment. After applying the above algorithm, the two original sequences become "WANDER-" and "WA-DERS". In this case, API sequences *Vs* and *Vu* are inserted with some zeros to generate *Vs'* and *Vu'*, which have optimal alignment.

We apply these similarity functions on *Vs'* and *Vu'*: Cosine measure, Extended Jaccard measure, and the Pearson correlation measure are the popular measures of similarity for sequences. These similarity measures are represented below

- **Cosine similarity:** is a measure of similarity between two vectors of *n* dimensions by finding cosine measure of the angle between them.

$$\text{Cosine Similarity} = \left[\frac{Vs' . Vu'}{\|Vs'\|_2 * \|Vu'\|_2}\right]$$

$$\text{Where } \|V\|_p = \left[\sum_i |V_i|^p\right]^{1/p}$$

- **Extended Jaccard measure:** measures the degree of overlap between two sets and is computed as the ratio of the number of shared attributes of *Vs*' AND *Vu*' to the number possessed by *Vs*' OR *Vu*'.

$$\text{Extended Jaccard Measure}=\frac{Vs'.Vu'}{\|Vs'\|_2^2 + \|Vu'\|_2^2 - Vs'Vu'}$$

- **Pearson correlation:** Correlation gives the linear relationship between two variables. For a series of measurements of variables *Vs*' and *Vu*', Pearson correlation is given by the formula below.

$$\text{Pearson Correlation} = \frac{1}{2}\left[\frac{\left(Vs'.-\overline{Vs'}\right)\left(Vu'-\overline{Vu'}\right)}{\left\|Vs'-\overline{Vs'}\right\|_2 . \left\|Vu'-\overline{Vu'}\right\|_2} + 1\right]$$

In this experiment, we calculate the mean value of the three measures. For a particular measure between a virus signature and a suspicious binary file, $S^{(m)}(Vs'_i, Vu')$, this stands for the similarity measure between virus signatures i and a

Table 2 Similarity Analysis (Mean value of all the three similarity measures) of malware and benign/normal programs

Malware	Back-door. Hack-Defend er.c1. virus	Back-door.H ackDe-fend-er.c3. virus	Back-door.Tr ojan.c1 2.virus	Back-door.Tr ojan.c1 6.virus	Con-struc-tion.Kit .c1. virus	Con-struc-tion.Kit .c3. virus	Hack-tool.Rx hide.c1 .virus	Hack-tool.Rx hide.c2. virus	Hack-tool.c5. virus
Back-door.Hack Defend-er.c1.virus	100	73.74	36.62	31.49	15.6	40.71	6.98	14.23	25.67
Back-door.HackDe fend-er.c3.virus	52.02	100	29.46	24.36	33.54	33.54	67.6	54.16	33.54
Back-door.Trojan. c12.virus	25.39	32.51	100	38.67	94.59	72.72	57.38	54.12	72.7
Back-door.Trojan. c16.virus	40.72	36.61	86.32	100	89.45	89.46	74.08	63.58	88.39
Construc-tion.Kit.c1.vi rus	15.2	11.15	12.16	9.11	100	13.18	41.79	35.66	13.18
Construc-tion.Kit.c3.vi rus	27.43	30.47	84.14	45.83	93.57	100	66.77	68.55	85.17
Hack-tool.Rxhide.c 1.virus	8.09	9.11	9.11	5.05	10.12	10.12	100	23.36	10.12
Hack-tool.Rxhide.c 2.virus	17.24	18.24	28.43	20.28	30.48	30.47	39.65	100	30.47
Hack-tool.c5.virus	23.35	28.45	84.13	45.84	93.56	94.6	72.79	65.48	100
Normal Executables									
REGSVR32. EXE	7.09	11.13	8.11	7.08	11.14	7.1	7.1	7.09	9.12
actmovie.exe	18.25	18.25	19.28	17.23	29.44	23.33	23.33	23.33	21.32
appletview-er.exe	9.11	0	2.03	7.09	9.12	10.12	10.12	10.12	8.1
arp.exe	4.05	9.11	9.11	7.08	0.01	0.01	0.01	0.01	4.05
attrib.exe	0	0	0.01	0	0.01	0.01	0.01	0.01	0
extcheck.exe	9.11	0	2.03	7.09	9.12	10.12	10.12	10.12	8.1

Table 2 *(continued)*

iisreset.exe	8.1	9.11	7.08	5.06	0.01	7.09	7.08	7.08	8.1
ipsec-mon.exe	9.11	9.11	14.18	11.14	20.29	15.19	15.19	14.18	12.14
java.exe	9.11	0	2.03	7.09	9.12	10.12	10.12	10.12	8.1
ping.exe	6.07	7.09	7.1	10.13	16.21	12.15	12.15	12.15	6.07
rsh.exe	7.1	13.17	15.19	16.21	16.21	17.23	17.23	14.18	10.12
rundll32.exe	10.12	10.12	1.02	7.08	10.13	7.08	7.08	7.08	9.11

suspicious binary file. Our similarity report is generated by calculating the $S^{(m)}(Vs'_i, Vu')$ value for each virus signature in the signature database.

The index of the largest entry in the similarity report indicates the most possible virus the suspicious file is (a variant of). Table 2 shows the preliminary results of our recent experiments and their similarity values compared against most lethal malware and well known benign executables. The value over the diagonal of Table 2 shows 100% since the sample is compared with the signature of itself in the database.

4.2 Feature Selection

In our approach the executables are disassembled and their API calls are parsed. Later we calculate the term frequency of these APIs and use support vector machines to classify them.

The collected samples are disassembled to extract API sequences. If the executable is packed, we unpack it using our methodology and later the new unpacked executable is reparsed to extract the new set of APIs and calculate its frequency of occurrence. In this study we just look for APIs being executed and not at their sequence.

We used the traditional but popular weighting scheme [18] borrowed from the field of information retrieval, and also used TF-IDF (term frequency-inverse document frequency). However in our study Term Frequency (TF) works the best, which references a term (in our case, an API) that appears several times within a document (in our case, an executable) that is likely to be more important than a term that appears only once. However, a few terms may appear more frequently in longer documents despite the actual importance of that term in the document. In order to prevent a biased situation like this TF is usually normalized. TF of each API is calculated using the formula below.

$$TF\ (w_{ij}) = n_{ij}\ /\ |d_i| \qquad (1)$$

Where, n_{ij} is the number of occurrences of an API w_j in an executable d_i and $|d_i|$ is total number of all the APIs in the executable d_i.

Inverse Document Frequency (IDF) on the other hand indicates that an API which occurs in a few executables is likely to be a better discriminator than an API that appears in most or all executables. This notion is very intuitive in both information retrieval and text categorization.

$$\text{IDF}(w_j) = \log(n / n_j) \tag{2}$$

Where, n is the total number of executables in the dataset and n_j is total number of executables where the API w_j appears. Multiplying TF with IDF we get the weight X_{ij} of a feature term (API) w_j in an executable i as:

$$X_{ij} = \text{TF}(w_{ij}) * \text{IDF}(w_j) \tag{3}$$

As a result, TF-IDF filters out the common APIs by giving a high weight to the API having high term frequency (in the given executable) and a low executable frequency to the API in the whole dataset.

The PE file is parsed so as to mine the data. Then the API sequences are extracted and each of its term frequencies is calculated. The value is then input into the classifier to identify whether the PE file is a malware or not. We use SVM for classification. From this methodology we achieved approximately 90% training accuracy and over 91.5% testing accuracy.

4.3 Experiments and Results

In this experiment the dataset contains 6038 executables, of which 3019 are malicious and the remaining 3019 are benign. The malware samples were collected from the offensive computing website [13].

In these data samples 3019 malware executables contain a random collection of viruses, Trojans, adware, spyware and a few of its variants. The data samples of 3019 benign executables are a collection of system files and normal executables. The data samples were split into 50%-50% for training and testing respectively. The training data set consists of 3019 data samples of which 1509 are malware executables and the rest are normal executables. Similarly the testing data set consists of 3019 data samples of which 1509 are from malware and the remaining are benign executables.

4.3.1 Model Selection of SVMs

In any predictive learning task, such as classification, both a model and a parameter estimation method should be selected in order to achieve a higher level of performance of the learning machine. Recent approaches allow a wide class of models of varying complexity to be chosen. The task of learning amounts to selecting the sought-after model; the one having optimal complexity and estimating parameters from the training data [19, 20].

Within the SVMs approach, usually, the parameters to be chosen are:

1. The penalty term C which determines the trade-off between the complexity of the decision function and the number of training examples misclassified;

2. The mapping function Φ; and
3. The kernel function such that

$$K(\mathbf{x}_i, \mathbf{x}_j) = \Phi(\mathbf{x}_i) \cdot \Phi(\mathbf{x}_j) \,.$$

In the case of RBF kernel, the width, which implicitly defines the high dimensional feature space, is the other parameter to be selected. Results using SVMs are given in Figure 12.

Figure 11 gives a model graph obtained during the training of SVM for malware dataset described in Section 4.3.

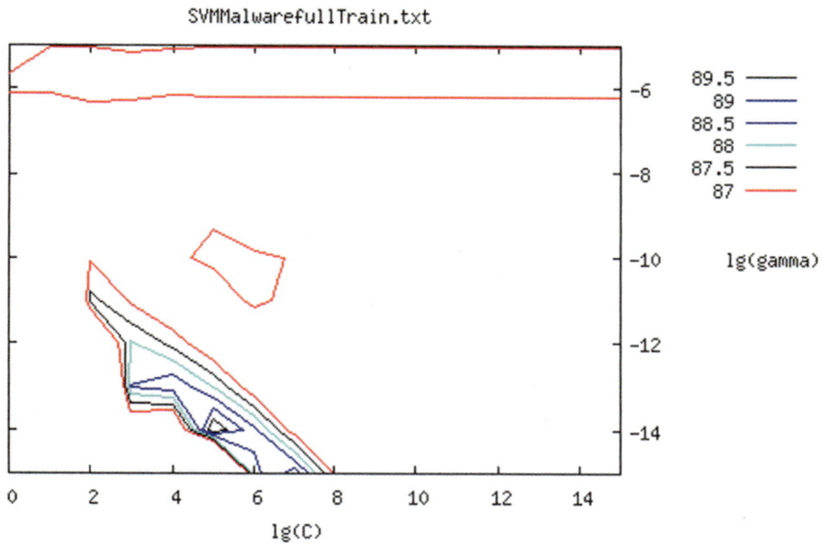

Fig. 11 SVM model graph for malware dataset

In Figure 11, the x-axis is term C and the y-axis is the kernel parameter γ (gamma). The graph is color coded based on the level of accuracy. The goal of model selection is to estimate the cost and the gamma parameter which produces best accuracy.

4.3.2 ROC Curves

Receiver Operating Characteristic (ROC) is a graphical plot between the sensitivity and specificity. The ROC is used to represent the plotting of the fraction of true positives (TP) versus the fraction of the false positives. The ROC curves are generated by considering the rate at which true positives accumulate versus the rate at which false positives accumulate with each one corresponding, to the vertical axis and the horizontal axis respectively as shown in Figure 12.

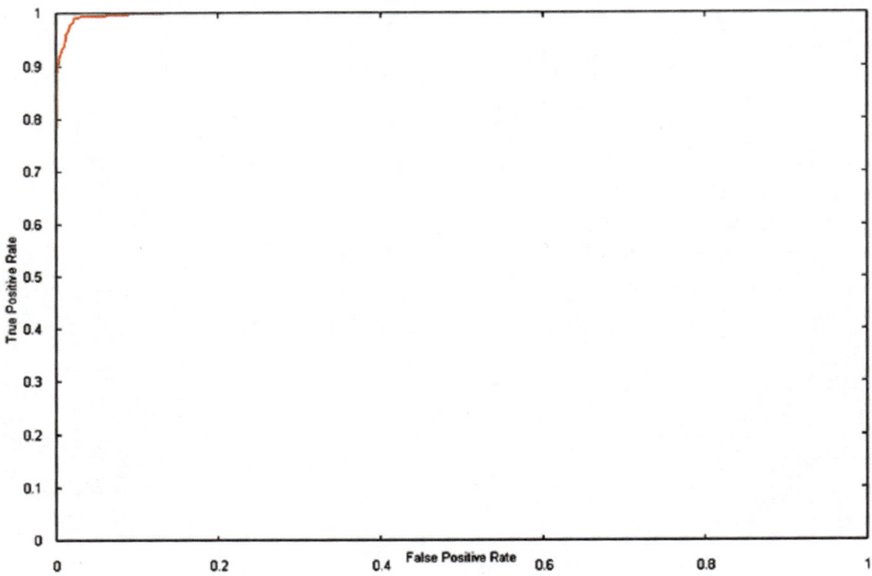

Fig. 12 Accuracy of SVM classification malware detection

The point (0, 1) is the perfect classifier, since it classifies all positive cases and negative cases correctly. Thus an ideal system will start by identifying all the positive examples and so that the curve will rise to (0, 1) immediately, having a zero rate of false positives, and then continue along to (1, 1).

Detection rates and false alarms are evaluated for the malware dataset described in section 4.3 and the obtained results are used to form the ROC curves. In each of these ROC plots, the x-axis is the false alarm rate, calculated as the percentage of normal executables detected as malware; the y-axis is the detection rate, calculated as the percentage of malware accurately detected (over 91.5% accuracy rate).

A data point in the upper left corner corresponds to optimal high performance, i.e., a high detection rate with a low false alarm rate [21]. Figure 12 gives the Receiver Operational Characteristics (ROC) curve generated by our malware detection model. From the graph it shows that even for 0% false positive rate, we achieved true a positive rate of 75% and for approximately 10% false positive rate, we achieved a true positive rate of 100%.

5 Summary and Conclusions

This chapter presents our approach for detecting malware based on its API calls. A few of our main contributions include:

- From the results of our experiments, we infer that we have accomplished our main goal of detecting malicious variants and malware that usually escape detection by commercial anti-virus in general.

- We evaluate our system based on large set of features (unique APIs), though the use of data samples is limited.
- Experiments performed on a large set of polymorphic malware showed that similarity measures are efficient in detecting polymorphic malware and variants of known malware.
- There is no necessity to study the sequence of API execution to classify a malware.
- Term frequency of an API is considered to classify an executable using SVM, one of the most widely used classifiers today.

After several iterative experiments, it is observed that SVMs easily achieve high detection accuracy (higher than 85%) for each of the datasets described in Section 4.3: malware dataset (please note real malware samples found over the Internet and benign samples were used in the experiments). MD5 hashes of the executables are used to verify their authenticity.

A grid search for malware detection using SVM (Figure 11) which seeks the optimal values of the constraint penalty for the appropriate solution model and the kernel width (C,γ), has been performed. We demonstrate that the ability with which SVMs can classify malicious code (malware) is highly dependent on both the kernel type and the parameter settings.

Based on the above mentioned observations we conclude that SVMs consistently achieve the best results for our dataset.

References

1. Symantec Threat Report January (accessed January 20, 2011), `http://www.symantec.com/content/en/us/enterprise/other_resources/bsymantec_report_on_attack_kits_and_malicious_websites_21169171_WP.en-us.pdf`
2. Nachenberg, C.: Computer virus-antivirus co-evolution. Communications of the ACM 40(1), 46–51 (1997)
3. Sanok Jr, D.J.: An analysis of how antivirus methodologies are utilized in protecting computers from malicious code. In: Information Security Curriculum Development (InfoSecCD) Conference, Kennesaw, GA, USA (1995)
4. Kephart, J., Arnold, W.: Automatic extraction of computer virus signatures. In: Proceedings of 4th Virus Bulletin International Conference, pp. 178–184 (1994)
5. Christodorescu, M., Jha, S.: Static analysis of executables to detect malicious patterns. In: Proceedings of the 12th USENIX Security Symposium (2003)
6. Rabek, J., Khazan, R., Lewandowski, S., Cunningham, R.: Detection of injected, dynamically generated, and obfuscated malicious code. In: Proceedings of ACM workshop on Rapid malcode, pp. 76–82 (2003)
7. Schultz, M., Eskin, E., Zadok, E.: Data mining methods for detection of new malicious executables. In: Proceedings of IEEE International Conference on Data Mining (2001)
8. Kolter, J., Maloof, M.: Learning to detect malicious executables in the wild. In: Proceedings of KDD 2004 (2004)

9. Yanfang, Y., Wang, D., Li, T., Ye, D.: IMDS: Intelligent Malware Detection System. In: Proceedings of KDD 2007 (2007)
10. Liu, B., Hsu, W., Ma, Y.: Integrating classification and association rule mining. In: Proceedings of KDD 1998 (1998)
11. Shen, Y., Yang, Q., Zhang, Z.: Objective-oriented utility-based association mining. In: Proceedings of IEEE International Conference on Data Mining (2002)
12. Provos, N., McNamee, D., Mavrommatis, P., Wang, K., Modadugu, N.: The Ghost In The Browser Analysis of Web-based Malware. Google, Inc. (2007)
13. Offensive Computing, `http://offensivecomputing.net` (accessed January 25, 2011)
14. Christodorescu, M., Kinder, J., Jha, S., Katzenbeisser, S., Veith, H.: Malware normalization. Technical Report 1539, University of Wisconsin, Madison, Wisconsin, USA (2005)
15. TechniZe Team, `http://www.technize.com/zeus-trojan-and-password-stealer-detection-and-removal` (accessed January 20, 2011)
16. Tarakanov, D.: `http://www.securelist.com/en/analysis/204792107/ZeuS_on_the_Hunt?print_mode=1` (accessed January 15, 2011)
17. Pietrek, M.: (2002), `http://msdn.microsoft.com/en-us/magazine/cc301805.aspx` (accessed January 8, 2011)
18. Salton, G., Wong, A., Yang, C.S.: A vector space model for automatic indexing. Communications of the ACM 18(11) (1975)
19. Cherkassy, V.: Model complexity control and statistical learning theory. Journal of Natural Computing 1, 109–133 (2002)
20. Lee, J.H., Lin, C.J.: Automatic model selection for support vector machines. Technical Report, Department of Computer Science and Information Engineering, National Taiwan University (2000)
21. Egan, J.P.: Signal detection theory and ROC analysis. Academic Press, New York (1975)

Intelligence Techniques in Computer Security and Forensics: At the Boundaries of Ethics and Law

Bernd Stahl, Moira Carroll-Mayer, David Elizondo, Kutoma Wakunuma, and Yingqin Zheng

Abstract. Computational Intelligence (CI) techniques have been widely used in the domains of computer security and computer forensics. One problem that normative discussions of technologies face is that the technical capabilities under investigation tend to be unclear and that the experts in normative questions do not tend to be experts in technical developments and vice versa. The present paper therefore sets out to chart the ethical and legal problems arising from a new and fast moving field, namely that of computational intelligence and its application to computer security and forensics. Using artificial neural networks (ANNs) as an example of computational intelligence, the paper's main aim is to create a link between what can now be perceived as technical developments and established discourses in ethics and the law. It aims to chart the territory to highlight likely ethical and legal problems related to ANNs and point in the direction of future research.

Bernd Stahl · Moira Carroll-Mayer · Kutoma Wakunuma
Centre for Computing and Social Responsibility, De Montfort University
Faculty of Technology
Department of Informatics
The Gateway, Leicester, LE1 9BH
United Kingdom
e-mail: {bstahl,mcm,kwakunuma}@dmu.ac.uk

David Elizondo
Cyber Security Centre, De Montfort University
Faculty of Technology
Department of Computing Technology
The Gateway, LE1 9BH, Leicester
United Kingdom
e-mail: elizondo@dmu.ac.uk

Yingqin Zheng
School of Management
Royal Holloway, University of London
Egham Hill, Surrey, TW20 0EX, United Kingdom
e-mail: yingqin.zheng@gmail.com

D.A. Elizondo et al. (Eds.): Computational Intelligence for Privacy and Security, SCI 394, pp. 237–258.
springerlink.com

1 Introduction

TECHNOLOGIES that relate to security and, maybe even more so those related to forensics have the potential to greatly help people and improve their lives. At the same time such technologies can be used for purposes of oppression and subjugation. Computational Intelligence (CI) techniques, such as artificial neural networks (ANNs), evolutionary computing and fuzzy logic, have been widely used in the domains of computer security and computer forensics.

Computer security concerns the protection of information and property from misuse by unauthorised individuals while making it accessible to the intended users. Computer forensics provides methods towards the preservation, collection, validation, identification, analysis, interpretation, documentation and presentation of digital evidence derived from digital sources for the reconstruction of criminal operations [1]. This paper reflects on issues arising from the use of CI in the security and forensics computer domains. To do this, the domain of ANNs is used as an example of CI techniques.

Ethics and the law are two different but interrelated ways for groups and societies to regulate behaviour. Behaviour is also affected in different ways by new technologies. There are several discourses covering general aspects of this relationship between normative disciplines (i.e. ethics, law) with technology and more particularly with computing-related technologies. These discourses include computer ethics, information ethics and computer law.

The contribution of this paper is to combine a number of disparate discourses and literatures. This will contribute to a better understanding of potential use of new technologies and thereby ensure that substantial ethical and legal problems can be avoided or overcome. However, the discourses involved refer to different disciplines and therefore to different understandings of what it means to do

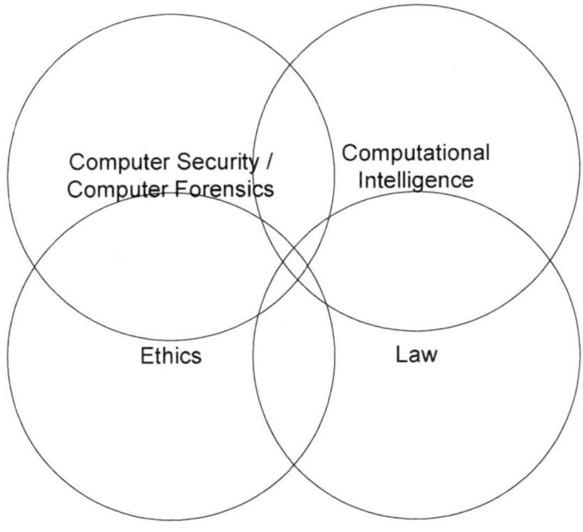

Fig. 1 Disciplines involved in analysis

research and how research problems are to be addressed. A schematic expression of the disciplinary complexity of the paper is represented in the following diagram:

The figure shows that it would be an improbable and overly ambitious endeavour to cover all of these areas, each of which is extensive in its own right. The purpose of the present paper is therefore to look at the overlaps, the cross section and explore which problems of the application of CI in computer security or forensics are predictable.

In order to develop this analysis, the paper will start by giving a brief description of CI and its related ANNs. Thereafter, it will discuss the technical side, that is the application of ANNs in the areas of computer security and forensics. It will then cover ethical issues that are foreseeable as a result of this and, in a further section, undertake a parallel analysis of the legal issues likely to arise. This will lead to a synthesis and discussion of normative issues. On the basis of this synthesis the paper explores ways of addressing these problems with a view to engage individual researchers and practitioners but also policy makers on possible courses of action.

2 Technology Description

Computational Intelligence can be said to be modelled on human-like intelligence where the technology is expected to perform functions that are at par or beyond human capability such as performing complex algorithm functions. As one example of CI, Artificial Neural Networks are modelled on and are an emulation of the biological human neural system. This means that, like humans, the expectation from ANNs is that such systems can be in control of human tasks, even go beyond human capability as well as emulate and carry out human-like functions. Application areas will include among others data mining, image recognition, speech recognition and even perform predictions [35]. In relation to data mining, Krieger [37] quoting Kosko [40] points out that Neural networks can be used in various ways and "programmed or trained to . . . store, recognize, and associatively retrieve patterns or database entries; to solve combinatorial optimization problems; to filter noise from measurement data; to control ill- defined problems; in summary, to estimate sampled functions when we do not know the form of the functions."

With ANNs as part of the research domain of Computational Intelligence, one of their main characteristics is that they can learn how to approximate functions and classify data from examples. A neural network can be seen as a computer system that is made up of simple and highly interconnected processing elements [11, 12]. These elements process information by their dynamic state response to inputs. Problems which were not solvable by traditional algorithmic approaches frequently can be solved using neural networks.

3 Uses of Neural Networks in Computer Security and Computer Forensics

3.1 ANN Applications in Computer Security

Examples of the use of neural networks in the field of security include:

1) Intrusion Detection: A growing number of computer systems are being compromised by intrusions. This explains the interest in building more robust systems for detecting intrusions. Neural networks have been developed that can detect Denial of Service and probing attacks [2]. Self organizing maps have also been used for intrusion detection.

2) Detection of Infected Computer Systems: Computers can be infected with viruses and Trojans. A neural network to detect zombie PCs was developed by [3]. Once the security is compromised and a virus enters the system, the zombie PC can be controlled remotely by the issuer of the virus. The system was capable of automatically detecting breached and infected computers.

3) Steganography: This is a technique used to hide information within other media. Watermarking is used as a means for preserving copyright digital information by embedding information into a digital document. The authors of [4] describe a neural network for watermarking documents. They used a chaotic sequence generator based on coupled Chebiscev maps combined with a back propagation neural network to develop a robust and secure watermarking system. A method for detecting digital media containing hidden information was developed in [5]. The system is capable of identifying jpeg files containing hidden information.

4) *User Authentication:* A neural network system for helping users to choose strong passwords was developed in [6]. The model checks passwords looking for features such as use of digits, strong and week passwords, upper and lower case characters, etc. Another model was developed by [7] that will authenticate passwords. The system prevents attackers from accessing and changing the list of stored passwords.

3.2 ANN Applications in Computer Forensics

Applications of neural networks in computer forensics cover:

1) *Computer Usage*: A neural network model for finding suspicious behaviour patterns during and analysis of temporary Internet files (which capture data about browsing history including sites visited and viewing durations) in four different computers used by people with similar work tasks was developed in [8].

Another model was developed in [9] for digital forensic use. The model aims to help locate relevant text quickly and accurately by clustering search results once they have been retrieved.

2) *Digital Evidence:* Neural networks have also been used to detect illegal activities including financial fraud using business related risk factors as inputs and a classification of fraud or not fraud output [10].

These are some indicative examples of the use of ANNs in computer security and forensics. The brief list does not claim to be comprehensive but gives some insight into typical applications that are currently being developed. For the purposes of this paper, they provide a starting point to identify how these new developments link to extant work on ethics and the law relating to Information and Communication Technologies (ICTs).

4 Ethics and CI in Computer Security / Forensics

Ethics can be seen as part of the societal background in which technology is developed and used. Ethics pertains to the society we collectively want to live in, the activities we find good or bad and acceptable or unacceptable. It is related to moral norms but also to the evaluation and justification of such norms. It has to do with morality and values. Quinn [13, p. 55] states that ethics is "a rational examination into people's moral beliefs and behaviour. Extrapolating this to ICTs, ethics relates to moral issues emanating from the development as well as use of ICTs. This will include how the ethical issues may affect users' lives but also larger societal consequences of ICT use. The expectation therefore, is that morals and values will be reflected in any technological development process and eventual technologies that are intended to add value and improve user's lives. This means that the technologies should be developed and used for the good of the intended user. As such, any technological systems need to take into account any potential difficulties that may arise during the course of their development and use. There is a variety of ethical theories and ways of linking them to ICT [14, 15, 16].. One way of organising the ethical review of CI in computer security and forensics is to consider the stakeholders who are likely to be influenced by such technologies. These include:

- Corporate computer end users (employees of organisations that use the technologies)
- Technology specialists (forensic analysts, security specialists)
- Organisations using the technologies
- Law enforcement agencies and officers
- Private computer users

There are many other stakeholders potentially involved in the use of such technologies but the above collection should suffice to show that there are differing interests and that they can lead to moral issues and problems of ethical evaluation. In addition, with different stakeholders with different interests come potentially different ethical issues for each group.

In order to further classify potential moral issues and provide a balanced view, we first discuss the benefits of the applications and then move to their disadvantages. This will be followed by an exploration of the ethical issues in general and as they may relate to each stakeholder group listed above.

4.1 Application Benefits

The above outline of the use of ANNs in computer security and forensics have given an indication of the benefits these technologies are likely to have. These include:

1. Increased security (i.e. reliability, accessibility) of data
2. Avoidance of access to data by unauthorised users
3. Avoidance of unauthorised change of data
4. Improved ability to identify users of data
5. Improved ability to detect breaches of the law

These are primarily points related to technical functioning but it is easy to see that they have ethical implications. Individuals usually desire their data to be secure. There is a strong link between information security and security overall, which is an important good with moral qualities [17]. Similarly, the capacities of ANNs in computer forensics have ethical implications in that they make it more likely that crimes are detected, prosecuted and punished. This links to society's desire to use law enforcement as a means of creating a better society, one in which individuals can live free from fear of being victims of crime.

By providing computational mechanisms of discharging these tasks, there is an added benefit of potentially reduced crime and better security at a lower cost and with higher accuracy. It is easy to see that a successful use of CI, for example in identifying illegal pictures, frees capacities for other law enforcement activities. It also has the advantage of greater speed and accuracy than human operators may have.

One can thus state that there are substantial moral benefits stemming from such technologies which support the development and use of such technologies. The question at the heart of this section is, however, how these advantages compare to potential disadvantages.

4.2 Moral Problems Arising from the Applications in General

Computer security and forensics technologies in general are very powerful technologies. They often work in ways that ordinary users do not understand and are not aware of. Some of the obvious problems arising from such powerful technologies are:

- *Data protection and privacy*: While security is of relevance to safeguard privacy, the powerful security and forensics technology contain the potential to do the opposite. Indeed, in the case of forensics technology, the very point of its application is to render data visible that users want to hide (e.g. evidence of criminal wrongdoing). It is thus not surprising that the concept of privacy is frequently named as the main moral concern in conjunction with computer security [18, 19].

- *Data Management and Trust*: This will have a direct relation to the above in that where data protection and privacy are not guaranteed; the user's expectation of good and robust management of their data becomes questionable. Users often use technology on the premise of trust, believing that their data is well managed and secure and not subject to for example forensic scrutiny regardless of wrongdoing or not. Of course, when it is the former one might argue that it is right to subject users to forensic scrutiny in order to curb crime such as online fraud. However, when innocent users are subjected to this, there is a moral dilemma that arises especially in terms of privacy as well as one's right to the control of one's data without fear of intrusion. Cavoukian [38] discusses how user's self-determination may be under threat if they experience loss of control.
- *Surveillance*: Partly related to privacy, but in many aspects going beyond it, is the question of surveillance. Surveillance is morally relevant, not only because it may breach privacy preferences but because it has psychological and social consequences. For example, social classification, stigmatization and exclusion are the potential effects surveillance may have on society, and often the people affected are not aware of this factor and its complicated roots [20]. The negative consequences of surveillance are often exacerbated by existing inequalities embedded in power relationships which is our next point [21].
- *Power relationships*: Given the powerful affordances of security and forensics technologies, ownership of such technologies can change existing power relationships. An employer, for example who uses ANNs to detect malicious changes to networks and configurations may easily change these tools to detect employee activities that they deem undesirable, notwithstanding the moral or legal evaluation of such activities. This raises the important question of who gets to use such technologies on which grounds and for which purposes. The current approach to leave the answer to this question to market forces appears ethically questionable [22].
- *Democratic development*: Given the preceding sections, one can argue that one important way of avoiding moral problems would be to involve users and stakeholders in general in the design of such technologies. There is, however, little evidence of the inclusion of stakeholder views in the design of such technologies. This raises the question of social and ethical description of definitions and assumptions that influence the development process. Given that these technologies are in many cases specifically created for the purposes of law enforcement, this furthermore raises the question of democratic overview of their implementations and implications. The current lack of such democratic oversight can be interpreted as an ethical issue in its own right.

All of the points enumerated above would be valid for most security and forensics technologies, whether they are digital or not. However, there are additional issues arising from the nature of the technologies analysed here. The principle of CI is to automate aspects of human intelligence. It aims to provide machines with the ability to make human judgments. The examples discussed above give a clear indication of the reach of the technologies. They are designed to make judgments concerning the nature of computer use, data provenance, and moral / legal acceptability.

In most cases the intention is to render daily tasks more manageable for human users. An intrusion detection system, for example, will typically alert a human network specialist to investigate whether an intrusion did happen. A forensic system used to classify files may alert a police officer of the potential presence of indecent images of children. An interesting ethical question arises from the extent of these activities and the human involvement in them. Will the overworked digital investigator rely solely on the system to determine the presence of illegal material? Will the network administrator leave the task of detecting anomalies to the system, possibly because she has no way of second guessing it?

The latter subsequently raises the question of loss of autonomy and freedom to be able to be in control of technologies like ANNs due to the fact that there may be a growing reliance on such technologies by humans. This is more-so as the technologies undertake more and more complex tasks and by implication decisions that used to be the preserve of human agents. As this becomes the norm, it may be worrying particularly when something goes wrong with the technology or when certain undertaken functions are not fully understood by the user and/or operator. This would suggest a relegation of autonomy or freedom on the part of the human agent to the technology. Bearing in mind that ANNs is modelled to automate human intelligence and therefore has control over certain and more complex tasks, the question is who should be held responsible and accountable in the event of potential harm resulting from the technology. Gill [39] argues that autonomy may not only be the preserve of humans especially when we consider how intelligent technological systems are used interactively with and alongside humans. With such an observation, the question of ethics continues to remain a concern.

In summarising the ethical issues in relation to the different stakeholders listed in one of the sections above, it would appear that for:

Corporate computer end-users: One of the usages of the technology would be for client's data. In this respect the technology is advantageous in that it can be used to hold huge amounts of client data as well as aid corporate end-users to perform complex tasks. However, ethical concerns will be how and whether client data is well managed, how possible technological difficulties may be rectified and the issue of privacy and trust in as far as client data is concerned.

Technologies specialists (forensic analysts, security specialists): ANNs will be particularly useful for this group of stakeholders in their duty to track criminal wrongdoing. As users leave a trail of information gathered when they use ICTs like the internet which can be identified by unique technology identifiers [38], it is easier for forensic analysts to pick up on such trails. Although this may be advantageous for security and law enforcement, it raises issues of surveillance, monitoring and invasion of privacy, particularly when innocent users are surveilled.

Organisations using the technology: As not everyone may understand how ANNs work even in organisation that may be inclined to use the technology, it prompts the question of knowledge and how limited or lack of robust understanding might impact others within and beyond organisations using the technology.

Law enforcement agencies and officers: In their quest to enforce the law, law enforcement officers might use the technologies in a way that may not be appropriate such as surveillance and an unwanted monitoring of users.

Private computer users: Often private computer users may not be aware that usage of ICTs leaves a trail of information which may be used to track and surveil them. This surveillance may be done by law enforcement officers, corporations or even fraudsters with the requisite know-how. Private computer users' data may be open to abuse or be misused in ways users have no knowledge of. It also raises issues of human rights and dignity as well as one's right to privacy if these cannot be guaranteed due to the nature of what the technology is capable of achieving.

These are questions of the efficacy of the use of technology and also how technology may operate based on its automated aspects of human intelligence. But they are furthermore questions of the place of human beings in the world and that of the technologies themselves. Early computing experts such as Wiener [23] and Weizenbaum [24] have asked the question what we can legitimately delegate to computers. This is a question that, in the light of the affordances of technologies such as the ones in question, needs to be revisited.

This section has discussed some of the ethical concerns arising from the use of CI in computer security and forensics. It is important to note, however, that these are not the only normative questions to be considered. Legal issues arise that complement the questions about the acceptance and acceptability of these technologies. These are discussed in the following section.

5 Foundational Impediments to the Reconciliation of Law and Security / Forensic Technologies

Guarantees of human rights in key instruments such as the European Convention on Human Rights (ECHR) and the Constitution of the United States are persistently undermined by the determination of legislators, legal practitioners, scientists and myriad experts to rely blindly upon technologies for forensics and security. Their reliance is unmatched by understanding of, or attempts to understand the actual propensities and implications of the technologies while misunderstanding at the interdisciplinary boundaries grows with technological complexity. Even within the scientific communities calls are for the greater understanding of inconsistencies of forensics capabilities with technological advancement; Al –Zarouni declares that '*Digital forensics relating to mobile devices seems to be at a standstill or evolving slowly*' [46] What is required according to Al Zarouni is an in depth framework for evaluating toolkits and data on appropriate tools and techniques for each type of phone.

Interrogation of the issues is also impeded by a vernacular which generally explicates the relationship between technology and law through counterpoising advanced technologies with legal stasis for example, Vacca (2005) [50] and Bainbridge (2009:239) [51]. This blaming of laws diverts the focus from the incompatibility of forensics and security tools with the capabilities ascribed to them by the scientific and legal professions.

A denouement is further delayed by indecision on whether activities in those domains constitute science or art or are inextricably entwined [49]. These trends conspire to deflect attention from the ability of forensics and security technologies to reconcile with legal standards for the protection of human rights in judicial and extra judicial settings.

6 Interdisciplinary Dissonance

Scientists, legislators, experts and lawyers lack the shared language or understanding to engage in normative discussions of the anomalies at the intersection of computer technologies, security, forensics and law. Omenn [52] notes a clash of cultures and language between scientists and lawyers '*Courtroom jargon is unfamiliar to scientists, just as scientific jargon is unfamiliar to the general legal community*'. The vastly more complex technologies for forensics and security prevalent in the 21st century reinforce the divergence noted by Omenn and instantiate under acknowledged technological anomalies in forensic and security technologies. Consequentially lawyers and scientists ascribe to the technologies capabilities unmatched by reality so that evidence obtained and its admissibility and acceptability in court and beyond is questionable. However attempts to dispel the dissonance impeding interdisciplinary understanding of the actual capabilities of technologies for forensics and security may stand or fall on whether activities in those domains constitute science or art or are inextricably entwined. The court in *United States v. Brooks*, 427 F.3d 1246, 1252 (10th Cir. 2005), conflates the fields to sustain a questionable digital search methodology, 'Given the numerous ways information is stored on a computer, openly and surreptitiously, a search can be as much an art as a science.' Carroll [49] rejects the need for argument, 'tools and methods are scientific and are verified scientifically, but their use necessarily involves elements of ability, judgment, and interpretation. Hence, the word " technique" is often used to sidestep the unproductive science/art dispute'. The designation of 'technique' as a place to which 'unproductive' disputes concerning science and art in digital forensics are assigned surely diminishes the concept of 'technique' in forensics and security as a locus of interdisciplinary understanding.

The problem of interdisciplinary dissonance can partially be ascribed to the 'silent' nature of security and forensics technologies which, according to Introna and Wood [25] (see Table 1), differentiate them qualitatively from other salient technologies.

Table 1. Silent and Salient Technologies [25, p.183]

Silent Technology
Embedded/hidden
Passive operation (limited user involvement)
Application flexibility (open ended)
Obscure (form/operation/outcome)
Mobile (software)

Salient Technology
On the 'surface'/conspicuous
Active operation (fair user involvement)
Application stability (firm)
Transparent (form/operation/hardware)
Located in hardware

The characteristics of technologies routinely used to gather and demonstrate forensic evidence for judicial purposes and the vastly more complex ubiquitous Artificial Neural Networks (ANNS), frequently employed in security analysis, are such that they qualify for categorisation as silent technologies as defined by Introna and Wood [25]. Their capabilities, imbedded into existing technologies make their operation impossible to detect, they require neither participation nor consent from subjects whose lives they affect, and involve, *prima facie*, a non-intrusive, contact-free process. Most pertinently the operation of ANNS is obscured by the inscrutability of software code in operation, electrical currents and untraceable instructions and executions [34] and [45].

A satisfactory response to the question posed by Peisert and Bishop [26] 'How fast would the car have had to be going in order to be crumpled like this?' extrapolated to forensics and security technologies demonstrates how non-salience precludes the certainty of questions and answers associated with more prosaic fields such as accident investigation. Peisert and Bishop's question elicits observable, finite calculations not the impossible story line of what cannot be humanly understood.

7 Reliability of Tools for Computer Security and Forensics

Technologies for security and forensics may be applied in the expectation of outcomes beyond their actual capabilities, the results obtained insufficient to properly implicate suspects. Operation Ore resulted in the suicides of numerous suspects following investigations, cautions and prosecutions on the basis of over simplistic interpretations of forensic computer evidence by law enforcement agencies. Conviction in the US, save for an offence of strict liability, requires the prosecution to bear the burden of proof of guilt beyond reasonable doubt as viewed by the reasonable man. In the UK the burden is similarly discharged or discharged where the jury is 'sure that the defendant is guilty' (see R v Majid [2009] EWCA Crim 2563). Forensic tools are frequently the sole source of evidence; for conviction in on-line child abuse cases, for example, they should uncover events leading up to the creation, viewing, deletion or modification of images, and link them to the defendant beyond reasonable doubt or in the UK so that the jury is sure the defendant is guilty. Interventions through pop-ups, spam, malware or fraud tending to disassociate the defendant from the charge must be accounted for too.

NIST [27] identifies numerous inaccuracies in forensics tools commonly used for investigating mobile phones which undermine the validity of their role in the evidential process including:

- Information could not be acquired from certain SIMs through the supported interface
- The remaining number of CHV attempts was not provided
- Data was rendered inconsistently in displays and reports
- Recovered data entries were truncated when displayed
- Recovered numeric data and English characters were not always decoded or translated correctly
- European and Asian character sets used in phonebook and message entries were not properly decoded
- Certain EMS messages were completely missed or their content unable to be recognized and rendered correctly
- New versions of a tool occasionally failed to perform as well as a previous one.

These issues are in addition to other research findings; mobile forensic toolkits are developed by third party companies and toolkits are not independently verified or tested for soundness. Developers use manufacturers' and self developed commands and access methods to gain data access to memory on devices [47]. Tools often limit themselves to one or more phone manufacturer handsets with a limited number of devices supported. Some tools are limited regarding connectivity options for the acquisition of data from handsets, for example, to wire connections as opposed to Infrared (IrDA) and Blue Tooth.

In addition some toolkits provide acquisition capabilities but not examination or reporting facilities [47]. Furthermore, mobile phone acquisition tools that claim to be forensically sound do not directly access memory but use commands provided by the phone's software and/or hardware interfaces for memory access thus relying on the forensic soundness of such software or hardware access methods [46]. Consequently when using these tools the ability to extract information without significantly changing the mobile phone's memory is not verifiable. Enhanced processor architecture in smart mobile phone devices using System on Chip (SoC) technology to incorporate distinct functionalities in the same package, reduce the number of chips required by it as well as incorporating built in memory further undermine the efficacy of mobile forensics [47].

However NIST [27] refuses to identify failing forensics tools thereby depriving disciplines of the very information that might enable understanding and removal of barriers to the correlation of computer forensics and security with fulfilment of laws designed to protect human rights judicially and extra-judicially.

That is not the sole reason for implicating NIST in the failure of disciplines to address the reconciliation of technologies for forensics and security with laws for the protection of rights. Some responsibility for the less than rigorous approach taken to the fitness of purpose of technologies for forensics and security in the judicial and non-judicial settings stems from uncritical reliance upon output from NIST concerning the technologies. Crucially there is an unfounded conflation of consideration of technologies by NIST with validation of those technologies due to an assumption that NIST concerns itself with validation of the technologies where in fact it does not. In the small print of its reports NIST strenuously denies any such role, for example,

'*Certain commercial entities, equipment, or materials may be identified in this document in order to describe an experimental procedure or concept adequately. Such identification is not intended to imply recommendation or endorsement by the National Institute of Standards and Technology, nor is it intended to imply that the entities, materials, or equipment are necessarily the best available for the purpose.*' [27] and [28].

Only by entering the search term 'disclaimer' on the NIST home page does the general disclaimer of any such role appear, 'Any mention of commercial products within NIST web pages is for information only; it does not imply recommendation or endorsement by NIST' [29].

In essence then NIST does not provide metrics to determine how accurately a tool works; it simply determines whether a set of requirements are, or are not, met. The belief that consideration by NIST of technologies for security and forensics equates with fitness for purpose is perpetuated by florid industry literature obliquely associating NIST consideration of their technologies with validation and infallibility. The dynamic is discernable in the following example from the NEC Corporation of America, a major contributor to law enforcement and defence forensics and security technologies:

> '*NEC's best-in-industry biometrics algorithms have been validated by the National Institute of Standards and Technology (NIST) in a comprehensive range of tests, from small one-to-one verification all the way up to large-scale, high-volume identification matching. Working with NEC means having access to decades of biometrics expertise and proven biometrics technologies that ensure performance, accuracy and reliability*'[30].

NIST places a great deal of reliance upon software developers and users of forensics and security technologies to address the problems noted by its research and of others. Just how much trust can be invested in the inclination of experts to ameliorate failings must be assessed in light of the American Bar Association findings

[48] that experts use tools like Norton Ghost or NT Backup for collection instead of validated forensic tools like EnCase or Forensic Toolkit (FTK).

There is a lack of distinction between data *extraction* and data *analysis* in digital forensics [49] which appears to have repercussions in distinguishing and determining the proper roles of digital forensics tools. The capabilities of Coroner's Toolkit, Sleuth Kit and En Case extend no further than the examination of file systems, copying, automated capture of data though claims are made of analytical properties. Microsoft's Computer Online Forensic Evidence Extractor COFEE, launched in December 2009 attracts similar commentary [37]. UNIX system log entries and comparatives commonly used for forensic analysis were designed with the debugging needs of programmers and system administrators in mind, not those of computer forensics. The Sun Basic Security Module and its cross-platform successors' construction is based on high-level assumptions about what events are important to security, not for linking individuals to actions in answer to specified forensic questions such as who did what [26]. The reliability of data derived through computer forensics is difficult to check without relatively expensive deterministic replay therefore relevant information is seen just once. If an attack occurs between two captures of information, traces of the attack may have been removed by the time the second snapshot is taken. BackTracker collects data more reliably than some other tools without the requirement of predicting security vulnerabilities using logging function, system library and kernel calls. BackTracker uses previously recorded calls and assumptions about call dependencies to generate graphical traces of events that have a□ected or have been a□ected by the file or process input. This correlation of data is effective and efficient but suspicious files and process IDs are not easy to discover especially where analysis occurs long after an intrusion so that there is uncertainty over input. Identifying a file through BackTracker may uncover a graphical trace of the processes and network sockets that instantiate a suspicious file but the starting point in the process remains elusive. In addition BackTracker cannot illuminate what happens within the process since it is primarily directed at a process-level granularity. Employing BackTracker to answer questions within the forensic process may therefore be inappropriate, even though the tool may appear to be accurate [26]. Tools currently used to gather information for evidential purposes should continue to be used only to the extent that they are actually useful. Sommer [31] asserts that even the Daubert principles (discussed below) are limited in usefulness by being more applicable to single issues of novel scientific evidence than those affected by programmes such as EnCase, AccessData, I-Look, SMART and Autopsy which typically incorporate numerous new, frequently updated, unpublished and untested principles.

8 The Methodological and Ontological Divide

At the heart of issues to do with the ethical and legal implications of technologies the traditional stand-off between positivist and interpretive methodologies and ontologies has a part to play. The dominance of the empirical approach in IS (to which category forensics and security technologies surely belong) leads to criticism that relevance is often sacrificed to rigor (Bharadwaj) [53]. Peisert and Bishop [26] draw attention to the consequential stress between that expected of forensic computing tools and their capabilities by scientists whose world view is empirically based and that of lawyers whose enquiries and inferences are qualitatively informed. The stress is manifest too particularly in the divergence between affects and the maintenance of human rights where ANNs are employed for optimizing security discussed above.

The position is particularly noticeable with regard to technologies for optimizing security such as ANNs and helps to explain the ascendency of AANs in roles such as that portrayed by Carley [35] and [36]. Since September 11th states, more intensely preoccupied with the perceived increased likelihood of attack, veer towards offensive realism, maximizing the technological capabilities domestically and internationally absent correlating interest in the implications for individual human rights. The twinning of realism with the efficiencies of maximized technological capabilities is spreading inexorably across technologically advanced states. Biddle [54] remarks,

> *'Realists see systemic [technological] competition as the key to international politics and expect states to converge on something like best practice under this competitive pressure...Realism thus predicts rapid and comprehensive diffusion of military methods with all quickly emulating the state of the art [technologies] as demonstrated by the strongest'.*

The preponderance of the positivist/offensive realist stance has important ethical and legal implications too for forensics technologies. As noted the divide between positivism and interpretive ontologies becomes apparent in judicial fora. Sommer [31] rightly asserts that the courts are not required to accept the scientific expert evidence proffered. It is the judges who are the gatekeepers whether on weight, favoured by UK courts or under the more specifically scientific grounds of the US Daubert requirements. The admissibility of expert evidence, such as that obtained through digital forensics tools, in UK courts is subject to the 'ordinary test of relevance and reliability' Rv. Bonython (1984) 38 SASR 45. However there is no consistent guidance on how reliability is to be determined as sufficient [55]. US courts employ the test described in Daubert v Merrell Dow requiring that for a forensics tool and the evidence it elicits to be admissible the following must be shown:

1. That the technique is subjected to peer review and publication.
2. The known or potential error rate of the technique.
3. The existence and maintenance of standards and controls concerning its operation.
4. Degree to which the theory and technique is generally accepted by a relevant scientific community.

However, it's difficult to meet the criteria due to a lack of error rates for most of the software used by the forensic experts. Additionally, there are no standards in

the field or peer reviews of methods. Another issue is whether it's possible to say that the 'relevant scientific community' has accepted the forensics tool.
Neither the UK nor US legal stance is robust enough to inculcate a tradition that adequately interrogates the technological issues that leave technologies for forensics and security at the boundaries of ethics and law.

The intractability of the science and law positions erodes the normative potential attributed to the law if it is accepted that legal norms reference commitment. Delacroix [33] rejects Cartesian eschewal that 'oughts' have the potential to be norms endowed with commitment and calls for a 'self-conscious reaction' that extrapolates to the legal and scientific communities involved in authentication of forensics and security technologies. As Gardner [34] cited by Delacroix acknowledges, `law (unlike morality) is something that one needs (further) reasons to obey, or indeed to engage with. Legal norms answer to morality'. Scientists and lawyers collaborating in the development, authentication and use of forensic and security technologies require a mutual, self-conscious reaction in the manner suggested by Delacroix [33] and Gardener [34]. Pressing questions must be asked by each community of the other and genuine answers sought if we are to move from the dubious ethical and legal imposition of technologies for forensics and security towards adequate risk assessment, transparency and proper balancing of the rights of individuals in society. The mutual impermeability of science and law promises worse than the merely unsatisfactory prescriptive approach to technologies complained of by Robinson et al [32, p.2]. Nowhere is the mutual adequacy of science and law as guarantors of legitimate legal rights challenged more than in the face of ambient intelligence characterized by ANNs [20, p.7], The importance of the mutual self-conscious reaction may best be assessed in light of the following section describing the employment of ANNs for security and purposes.

9 Legal Implications of Forensic Data Mining through Artificial Neural Networks

Artificial Neural Networks (ANNs) stand at the apex of state of the art forensic and security capabilities. However, the technology does not necessarily bode well for the human rights of those affected by its employment in forensics and security. The ECHR and the UN Convention jointly and severally protect the right to life, a fair trial, respect for private and family life, freedom of thought, conscience and religion, of expression, to hold opinions and receive and impart information, freedom of peaceful assembly and association with others. There are indications that data mining and analysis through ANNs, for forensic and security purposes, enabling governments to learn about individuals from the data they generate, may be incompatible with these rights. Data mining and analysis through ANNs may open direct and oblique routes to stigmatisation, discrimination and false identity management through ostensibly objective and neutral data collection and analysis, stereotyping racial, social characteristics and behaviour. For example technologies 'flagging' disproportionate numbers of people with a middle-eastern appearance or those displaying unusual behaviour, as defined by computer systems, risk the misinterpretation of these characteristics as 'objective proof' of their posing a security risk or of identity mismanagement leading to loss of life [41]. ANNS represent efficiency led by mathematics rather than emotion and human judgment but they risk subordinating outcomes for humans to inscrutable and uncontrollable digital feedback loops. Combined algorithmic interaction in ANNS can create reactive conditions that are incomprehensible to the human mind and impossible to predict [45]. This conclusion, and its implications for security purposes, is borne out by Hurewitz and Marwala [37, p. 110] who comment on a typical ANN application for market analysis,

> 'In the field of machine learning, it is still remarkably difficult to implement reinforcement learning when using nonlinear function approximators, such as neural networks. This limitation makes application to nonstationary systems a challenging endeavor. A further limitation...is the inadequacy of traditional risk measurement benchmarks, as their [users] unrealistic assumptions taint the underlying analysis'

The irreconciliation of ANN capabilities with evaluation of non-stationary systems alluded to by Hurwitz and Marwala [37] is of concern since the fluidity of human cohorts, most particularly during societal upheaval or violence (the very moment ANNs might be relied upon for security enhancement) likely intensifies. The only way of evaluating ANNs capabilities is to examine their performance under controlled conditions; unfortunately appropriate control conditions are chiefly dominated and are under the auspices of US defence entities [25] rendering access difficult and somewhat compromising the neutrality of the controlled conditions.

Carley [35] and Carley and Behrman [36] indicates that forensic and security data obtained and analysed through ANNs may be associated with and encourage violations of the right to life through 'extra-judicial killings' invoked by the state or its agents and carried out without a fair trial. Carley [35, p. 182] presents conclusions, derived from an ANN analysis, conducted in 2004 [35], that the 'removal' of identified individuals weakens the cohesion of Hamas and Al Qaeda groups. Carley and Bhrman [36] describes how the ANN calculated 'an expected change over time' for Al Qaeda and Hamas; calculations were also made of the expected state '*if we just left them alone..., and what would happen if the top leader were removed from each. We did this analysis a while back when Yassin was the Hamas leader and, in fact, Hamas' performance did improve once Yassin was removed.*' Sheik Yassin was subsequently killed in a targeted attack by a US drone in 2004.

Interestingly Carley [35, p. 16] indicates that the ANNs analysis described in Carley and Behrman [36] was partly funded by the US Defense Advanced Research Projects Agency (DARPA). DARPA is the main funding and acquisitions partner in the development and fielding of Unmanned Combat Aerial Vehicles (Predator MQ-1/Reaper MQ-9 drones) one of which conducted the fatal strike on Sheik Yassin in 2004. Worryingly, since here one is talking about the human right to life among other rights, Carley describes the frequent error rate for ANNs of 10% to 20% as a 'kind of fidelity' [35, p. 187].

ANNs assisted data mining for security forensic analysis may erode other freedoms extended by the ECHR and the US Constitution. Carley [35, p. 170] describes ORA a statistical toolkit for meta-matrices that identifies vulnerabilities, key actors (including emergent leaders), and network characteristics of groups, teams, and organizations typical of those informing the Hamas/Al-Qaida analysis described. ORA runs a series of statistical tool kits on the networks with a set of measures organised into reports for intelligence, management and risks. ORA utilises multiple data sets; who-talks-to-whom, who has access to what resources, who has been involved in what events, who has been seen at what location. It utilises this data to predict, think about, and infer actions from one network to another. Patterns of connectivity based, for example, upon individuals sharing unspecified resources infer they have actually met. That assumption is then taken to infer many other social connections un-inferable without AANs [35, p. 173]. ORA views the sharing of resources as pivotal however the absence of descriptors of resources in [36] leaves open the possibility that they are, for example, water, oil, the services of a bar or hotel, and other ordinarily innocuous phenomena. Neither is it difficult to imagine as a consequence of ANN activities in forensics and security deliberate/enforced changes to individuals' normal behaviour and socialisation patterns synonymous with those protected by laws. The rights foregone might include the rights to assembly, expression, private and family life, thought, conscience, religion and opinion. Proponents in democratic society of ANN forensics and security exercises need to be aware that it is still remarkably difficult to implement reinforcement learning [37, p. 10], such as that inherent in ORA and other nonlinear neural networks. This limitation makes application to non-stationary systems such as populations or groups a challenging endeavour, calling into question their suitability at all for decisions of life and death. Added to this is the noted inadequacy of traditional risk measurement benchmarks for ANNs [32] and the presence of unrealistic assumptions that undermine the validity of ANNs' analyses [38].

10 Synthesis and Possible Solutions of Normative Issues

This paper has attempted to give an overview of normative issues arising from the use of CI technologies in computer security and forensics. The nature of this paper is exploratory, so it does not claim comprehensiveness of issues or their discussion. Having discussed the moral and ethical as well as the legal issues that arise in the context of CI in security / forensics, one can ask what the common theme is that emerges.

One can distinguish between specific / substantial issues, general problems of the topic area and those that relate to the specific properties of CI. The specific and substantial issues are those that arise due to the specifics of the technology. To go into more depth into any of them would require a more detailed description of the technology. Intrusion detection systems, for example, may have manifold uses and consequences and their specific ethical implications would need to be explored in conjunction with a specific discussion of the problems.

The next set of normative issues are those relating to security and forensics in general, which may take specific forms based on the use of CI. These might be privacy concerns or power issues on the ethical side or questions of due process on the legal side. None of these are unique to CI but they may be exacerbated due the use of novel technologies.

The most interesting normative issues seem to be those intimately related to the specific characteristics of CI, namely its capacity of emulating human intelligence. The potential autonomy of the technology raises a wide range of ethical as well as legal problems. There are empirical questions relating to the extent to which ANNs can replace human judgments, anthropological concerns about to which degree we want to allow technology to take human roles and legal questions regarding the way in which legal precedent can be applied to machines or in which way they limit the application of law.

An exploratory paper such as this cannot give answers to such complex questions. But it can try to outline a way that would allow addressing these issues simultaneously. The one problem that pervades all of the different normative issues is the loss of human control over consequences of technological agency. If this is an appropriate high level summary of the problems, then the solution is likely to lie somewhere in the area of governance of technology, its use and development, with a view to regaining this lost control.

What seems to be at issue is that technical developments are moved forward and implemented by individuals and organisations who do not necessarily consider the normative issues described here nor do they have an obvious moral or legal duty to do so. The consequences of such technical development, however, go far beyond the individuals or organisations that initiate them. The normative question arising from this is then how democratic control can be regained of such issues that are arguably a legitimate subject of democracy.

Again, the answer to this question is not trivial and will require efforts on many levels. These include the education of individual developers who will need to realise their professional responsibility to the greater good of society. Organisations need to consider their role in overall society. This will require the setting of governance and incentive structures that will allow individuals and organisations to act on a well-grounded view of the public good. Overall, this is part of the greater question how technology, its development and use should be treated in democratic societies. The present paper suggests that an important step in giving an answer to this is to develop awareness of the normative issues and hopes to have made a contribution to the answer to this question by outlining some of the issues that should be considered.

Acknowledgements: The research leading to these results has received funding from the European Community's Seventh Framework Programme (FP7/2007-2013) under grant agreement n° 230318.

References

[1] DFWRS, A road map for digital forensics research. Digital Forensics Research Workshop (August 2001), http://www.dfrws.org/2001/dfrws-rm-final.pdf

[2] Pachghare, V.K., Kulkarni, P., Nikam, D.M.: Intrusion Detection System using Self Organizing Maps. In: International Conference on Intelligent Agent & Multimedia-Agent Systems, IAMA (2009)
[3] Salvador, P., et al.: Framework for Zombie Detection Using Neural Networks. In: Fourth International Conference on Internet Monitoring and Protection, ICIMP (2009)
[4] Wei, P., Zhang, W., Yang, H., Huang, S.: Combining neural networks and chaotic map to construct blind digital watermarking scheme. In: Proceeding of the Sixth International Conference on Intelligent Systems Design and Applications, ISDA (2006)
[5] Lafferty, P., Ahmed, F.: Texture based steganalysis: results for color images. In: Mathematics of Data/Image Coding. Compression and Encryption VII. The Catholic University of America, 620 Michigan Avenue, NE, Washington DC, (2006)
[6] Ciaramella, A., D'Arco, P., De Santis, A., et al.: Neural Network Techniques for proac-tive passwords checking. IEEE Transactions on Dependable and Secure Computing 3(4), 327–339 (2006)
[7] Wang, S., Wang, H.: Password authentication using Hopfield neural networks. IEEE Transactions on Systems Man and Cybernetics – Part C: Applications and Reviews 38(2), 265–268 (2008)
[8] Fei, B.K.L., Eloff, J.H.P., Olivier, M.S., Venter, H.: The use of self-organizing maps of anomalous behaviour detection in a digital investigation. Forensic Science International 162, 33–37 (2006)
[9] Beebe, N.L., Clark, J.: Digital Forensic text string searching: Improving information retrieval effectiveness by thematically clustering search results. Digital Investigation 4S, S49–S54 (2007)
[10] Chen, H., Huang, S., Kuo, C.: Using the artificial neural networks to predict fraud litigation: Some empirical evidence from emerging markets. Expert Systems with Applications 36, 1478–1484 (2009)
[11] McClelland, J., Rumelhart, D.E.: And The PDP Research group, Parallel distributed processing: explorations in the microstructure of cognition. Foundations I (1986A)
[12] McClelland, J., Rumelhart, D.E.: And The PDP Research group, Parallel distributed processing: explorations in the microstructure of cognition. Psychological and Biological Models ll (1986B)
[13] Quinn, M.J.: Ethics for the Information Age, 2nd edn. Pearson International Edition (2006)
[14] Floridi, L. (ed.): The Cambridge Handbook of Information and Computer Ethics. Cambridge University Press (2010)
[15] Himma, K.E., Tavani, H.T. (eds.): The Handbook of Information and Computer Ethics. Wiley, Hoboken (2008)
[16] Hoven, J.V.D., Weckert, J. (eds.): Information Technology and Moral Philosophy. Cambridge University Press, Cambridge (2008)
[17] Brown, W.S.: Ontological Security, Existential Anxiety and Workplace Privacy. Journal of Business Ethics 23(1), 61–65 (2000)
[18] Conger, S., Landry, B.J.: The Intersection of Privacy and Security. Presented at the Association of Information Systems SIGSEC Workshop on Information Security & Privacy (WISP 2008), Paris (2008)
[19] Myyry, L., Siponen, M., Pahnila, S., Vartiainen, T., Vance, A.: What levels of moral reasoning and values explain adherence to information security rules? An empirical study. European Journal of Information Systems 18(2), 126–139 (2009)

[20] Lyon, D.: Surveillance society: Monitoring everyday life. Open University Press (2001)

[21] Introna, L.D.: Workplace Surveillance 'is' Unethical and Unfair [Opin-ion]. Surveillance & Society 1(2), 210–216 (2003)

[22] Stahl, B.C.: Forensic Computing in the Workplace: Hegemony, Ideology, and the Perfect Panopticon? Journal of Workplace Rights 13(2), 167–183 (2008)

[23] Wiener, N.: The human use of human beings. Doubleday (1954)

[24] Weizenbaum, J.: Computer Power and Human Reason. W. H. Freeman and Company, San Francisco (1976)

[25] Introna, L.D., Wood, D.: Picturing Algorithmic Surveillance: The Politics of Facial Recognition Systems. In: Norris, McCahill, Wood (eds.) Surveillance and Society CCTV Special, vol. 2(2/3) (2004)

[26] Peisert, S., Bishop, M.: Computer Forensics In Forensis (2007), `http://www.cs.ucdavis.edu/~peisert/research/2008-SADFE-ForensicsInForensis-PBM.pdf`

[27] Jansen, W., Delaitre, A.: NIST, Mobile Forensic Reference Materials: A Methodology and Reifica-tion (2009)

[28] Hu, V.C., et al.: NIST, Interagency Report 7316 Access to Access Control Systems (2006), `http://csrc.nist.gov/publications/nistir/7316/NISTIR-7316.pdf`

[29] NIST Homepage, `http://www.necam.com/IDS/Validation/`

[30] NEC Corporation of America (2011), NIST Validation, `http://www.necam.com/IDS/Validation/`

[31] Sommer, P.: Scientific, Technological and Expert Evidence in Law and Legal Proceedings. In: Artificial Intelligence and Forensic Computing Workshop (2009), `http://www.ktn.qinetiq-tim.net/content/files/groups/aiforensics/2009-04-02_PeterSommer.pdf`

[32] Robinson, et al: (RAND) Review of Data Protection Directive: Summary prepared for the Information Commisioner's (2009), `http://www.ico.gov.uk/upload/documents/library/data_protection/detailed_specialist_guides/review_of_eu_dp_directive_summary.pdf`

[33] Delacroix, S.: You'd better be committed: legal norms and normativity. American Journal of Jurisprudence (December 2009)

[34] Gardner, J.: Nearly natural law. American Journal of Jurisprudence 52(1), 1–24 (2007)

[35] Carley, K.: Dynamic Network Analysis in Counterterrorism Research. In: Proceedings of a Workshop on Statistics on Networks, Board of Mathematical Sciences (2007), `http://books.nap.edu/openbook.php?record_id=12083&page=169`

[36] Carley, K., Behrman, R.: Social Network Influences on Strategic Choices (2004), `http://www.casos.cs.cmu.edu/publications/papers/behrman_2004_socialnetworkinfluences.pdf`

[37] Hurwitz, E., Marwala, T.: State of the Art Review for Applying Computational Intelligence and Machine Learning Techniques to Portfolio Optimisation (2009)

[38] Kasabov, N.: Foundations of Neural Networks, Fuzzy Systems, and Knowledge Engineering. The MIT Press Cambridge, Massachusetts (1998)

[39] Kenneally, E.: Gatekeeping Out Of The Box: Open Source Software As A Mechanism To Assess Reliability For Digital Evidence. Virginia Journal of Law and Technology (Fall 2001), http://www.vjolt.net/vol6/issue3/v6i3-a13-Kenneally.html
[40] Krieger, C.: Neural Networks in Data Mining (1996), http://www.cs.uml.edu/~ckrieger/user/Neural_Networks.pdf (retrieved on June 28, 2010)
[41] Cavoukian, A.: Privacy in the clouds. Identity in Information Society (1), 89–108 (2008)
[42] Gill, S.: AI & Society 22, 265–270 (2008), http://people.na.infn.it/~tamburrini/pub/Guest%20editorial.pdf
[43] Kosko, Bart: Neural Networks and Fuzzy Systems. Prentice Hall, New Jersey (1992)
[44] Carroll-Mayer, M., Fairweather, B., Carsten Stahl, B.: CCTV Identity Management and Implications for Criminal Justice. Some Considerations in Surveillance and Society 5(1), 33–50 (2008) http://surveillance-and-society.org/articles/identity.pdf
[45] Salmon, F., Stokes, J.: Algorithms Take Control of Wall Street. In: WIRED (January 2011)
[46] Al-Zarouni, M.: Mobile Handset Forensic Evidence. A Challenge to Law Enforcement (2006)
[47] Jansen, W., Ayers, R.: Guidelines on Cell Phone Forensics (2006), http://csrc.nist.gov/publications/drafts/DraftSP800101
[48] Shirk, E.: The Dangers of Do-It-Yourself Computer Forensics. In: Law Practice Today, The American Bar Association (2007)
[49] Carroll, O. (2008) Computer Forensics: Digital Forensics Analysis and Methodology, US Attorneys' Bulletin (January 2007)
[50] Vacca, J.: Computer Forensics: Computer Crime Scene Investigation, vol. 1. Charles River Media (2005)
[51] Bainbridge, D.: Intellectual Property. Pearson (2009)
[52] Omenn, G.: Enhancing the Role of the Scientific Expert Witness 102(8) (August 1994)
[53] Bharadwaj, A.: Integrating Positivist and Interpretive Approaches to IS Research (2005)
[54] Stephen, B.: Review of The Diffusion of Military Technology and Ideas. In: Goldman, E., Eliason, L. (eds.) Perspectives on Politics, vol. 3(3). Stanford University Press, Stanford (2004)
[55] Leveson, L.J.: Expert Evidence in Criminal Courts. The Problem (2010)

Author Index